# Rings with Involution

**Chicago Lectures in Mathematics**
Irving Kaplansky, Editor

*The Theory of Sheaves,* by Richard G. Swan (1964)
*Topics in Ring Theory,* by I. N. Herstein (1969)
*Fields and Rings,* by Irving Kaplansky (1969; 2d ed. 1972)
*Infinite Abelian Group Theory,* by Phillip A. Griffith (1970)
*Topics in Operator Theory,* by Richard Beals (1971)
*Lie Algebras and Locally Compact Groups,* by Irving Kaplansky (1971)
*Several Complex Variables,* by Raghavan Narasimhan (1971)
*Torsion-Free Modules,* by Eben Matlis (1972)
*The Theory of Bernoulli Shifts,* by Paul Shields (1973)
*Stable Homotopy and Generalised Homology,* by J. F. Adams (1974)
*Banach Algebras,* by Richard D. Mosak (1975)
*Rings with Involution,* by I. N. Herstein (1976)
*The Theory of Unitary Group Representations,*
by George W. Mackey (1976)

Chicago Lectures in Mathematics



# rings with involution

I. N. Herstein



The University of Chicago Press
Chicago and London

I. N. HERSTEIN is professor of mathematics at the University of Chicago. His books include the widely used text *Topics in Algebra* as well as *Noncommutative Rings* and *Topics in Ring Theory* (1969).

The University of Chicago Press, Chicago 60637
The University of Chicago Press, Ltd., London

 Published 1976
Printed in the United States of America
80 79 78 77 76 987654321

*Library of Congress Cataloging in Publication Data*

Herstein, I N
Rings with involution.

(Chicago lectures in mathematics)
Includes bibliographies.
1. Rings with involution. I. Title. II. Series.
QA251.5.H49 512'.4
ISBN 0-226-32805-8
ISBN 0-226-32806-6 pbk.

To the memory of my mother

CONTENTS

# PREFACE

I have tried to give in this book a rather intense sampler of the work that has been done recently in the area of rings endowed with an involution. There has been a lot of work done on such rings lately, in a variety of directions. I have not attempted to give the last-minute results, but, instead, I have attempted to present those whose statements and proofs typify the kind of things that are being done.

The results from this area have important applications outside of associative ring theory itself. One such broad area of application is to the theory of Jordan algebras and of quadratic Jordan algebras. I have not developed this interrelationship here, not because it isn't interesting or relevant but because it would have taken me too far afield. It also would have expanded the book immensely. We recommend that the reader interested in this connection go to the mathematical literature, especially to the papers by N. Jacobson, K. McCrimmon, and M. Osborn, where such applications are readily found.

A second area of applications is that of operator algebras and Banach algebras. Here, too, for the reasons cited above, I decided not to develop these applications here. The interested reader can find such connections in the papers by Miers, Sunouchi, and Topping and in the Springer Lecture Notes by de la Harpe.

In writing this book, one source of great and constant pleasure has been the extent to which the work by my students plays a central role in almost everything that I do. These students are W. Baxter, C. Lanski, P.H. Lee, W. Martindale, S. Montgomery, Lynne Small, and M. Smith.

I am especially grateful to Prof. Susan Montgomery for all the help she has given me in bringing this project to an end. Not only do her results make frequent appearances throughout the book, but she has made many suggestions which have improved the proofs and the exposition. This is most apparent in the last part of Chapter 2 and in all of Chapter 5.

February 6, 1976
Chicago, Illinois

# 1

# RING-THEORETIC PRELIMINARIES

Our main objective in this book will be to describe some recent work in the general area of rings having an involution. Before we can get down to the discussion of these rings we must develop some results which are in the more general context of ring theory, for rings which do not necessarily have an involution.

There will be certain easy and formal facts about rings which we shall constantly have to use throughout the book. These results are, by and large, well-known and readily found. In order to have a ready reference when we shall use them, and for the sake of completness, we shall do these results in full detail in the first section of this chapter.

In the later sections of this chapter, the material becomes much less formal and cuts a great deal deeper.

In the second section we give a rather precise description of primitive rings, and of primitive rings with involution, which have a minimal one-sided ideal.

In the third section we derive some pretty and important theorems about rings satisfying polynomial identities and generalized polynomial identities. These results play a crucial role in much that will follow in the book.

In the fourth section we continue with the theme of polynomial identities. The main result is a lovely theorem due to Formanek which shows that in the $n \times n$ matrices over a field there are polynomials that take on non-constant values in the center. This result is now of extreme importance in all of ring theory; in our set-up of rings with involution it will also play an important role.

In the fifth section we give a remarkably beautiful and simple proof, recently given by Rosset, of the important Amitsur-Levitzki theorem which states that the $n \times n$ matrices over a commutative ring satisfy the standard identity of degree $2n$.

In the sixth, and final, section we obtain some theorems which intimately interrelate the nature of the centralizers of certain types of elements in a ring with the global structure of the ring itself.

In everything that follows, the word "ideal" will mean "two-sided ideal." Also, throughout the book we shall have occasion to call on results which we have proved in our little monograph *Topics in Ring Theory*, University of Chicago Press, 1969. When we cite these results we shall do so by referring to them as Lemma X or Theorem X in TRT.

## 1. Some Formal Results

We begin by defining two familiar kinds of rings which will come up almost everywhere in this work. All rings discussed will be associative, but we do not insist that they have unit elements.

DEFINITION. A ring $R$ is said to be prime if whenever $I_1 \neq 0$ and $I_2 \neq 0$ are ideals of $R$ then $I_1 I_2 \neq 0$.

As is virtually trivial, the definition above of primeness is equivalent to the statement: if $aRb = 0$, where $a, b \in R$, then either $a = 0$ or $b = 0$. Equally trivial is the fact that $R$ is prime if and only if the right-annihilator of a non-zero right ideal of $R$ must be $0$. It follows immediately from this that if $\lambda \neq 0$ is a left ideal, and $\rho \neq 0$ a right ideal, in the prime ring $R$ then $\lambda \cap \rho \neq 0$.

DEFINITION. A ring $R$ is said to be <u>semi-prime</u> if it has no non-zero nilpotent ideals.

Semi-primeness, like primeness, can be characterized in terms of elements of the ring. This characterization runs as follows: $R$ is semi-prime if and only if $aRa = 0$, with $a \in R$, forces $a = 0$.

We begin with an extremely easy and useful result.

LEMMA 1.1.1. If $R$ is a prime ring with no non-zero nilpotent elements then $R$ has no zero divisors.

<u>Proof.</u> Suppose that $ab = 0$; since $(ba)^2 = b(ab)a = 0$, by our hypothesis on $R$ we must conclude that $ba = 0$. However, if $ab = 0$ then $a(bx) = 0$ for all $x \in R$, whence, by the above, $bxa = 0$. Since $bRa = 0$ and $R$ is prime, we must have $a = 0$ or $b = 0$.

This result is capable of extension to products of more than two elements. In fact, to make this extension we need not work in rings, for it holds in the more general context of semi-groups.

LEMMA 1.1.2. Let $A$ be a multiplicative semi-group with $0$, and suppose that $A$ has no non-zero nilpotent elements. If $a_1, \ldots, a_n \in A$ and $a_1 a_2 \cdots a_n = 0$ then $a_{i_1} a_{i_2} \cdots a_{i_n} = 0$, where $i_1, \ldots, i_n$ is a permutation of $1, 2, \ldots, n$.

Proof. As in the proof of Lemma 1.1.1, if $ab = 0$ in A then $ba = 0$. Hence, if $ab = 0$ then $bax = 0$ for any $x \in A$, and so $axb = 0$.

From $a_1a_2 \cdots a_n = 0$, by the above remark we have that $a_na_1a_2 \cdots a_{n-1} = 0$. Thus the cyclic permutation $(1\ 2 \ldots n)$ is one for which the statement of the lemma is valid.

We claim that if $a_1a_2 \cdots a_n = 0$ then $a_2a_1a_3 \cdots a_n = 0$. For, by the remark of the first paragraph, using it successively we obtain $x_1a_1x_2a_2 \cdots a_{n-1}x_na_n = 0$ for any $x_1, \ldots, x_n \in A$. Let $x_1 = a_2$, $x_2 = a_3 \cdots a_n$, $x_3 = a_1$. This gives $(a_2a_1a_3 \cdots a_n)^2 = 0$, hence by hypothesis, $a_2a_1a_3 \cdots a_n = 0$. Thus the lemma is valid for the permutation $(1\ 2)$. Since $(1\ 2)$ and $(1\ 2 \ldots n)$ generate the symmetric group of degree n, we have prove the lemma.

DEFINITION. A ring R is called a domain if it has no zero divisors.

Note that we don't require that R be commutative in this definition of domain. We now can prove a pretty theorem due to Andrunakievitch and Rjahubin [5].

THEOREM 1.1.1. Let R be a ring with no nilpotent elements. Then R is a subdirect product of domains.

Proof. We first show: if P is a minimal prime ideal of R then R/P is a domain. For, let M be the complement of P and A the multiplicative semi-group of R generated by M.

We claim that $0 \notin A$. For, if $m_1m_2 \cdots m_k = 0$ where $m_1, \ldots, m_k \in A$, since P is a prime ideal, for some $x_1, \ldots, x_{k-1} \in R$, $m_1x_1m_2x_2 \cdots m_{k-1}x_{k-1}m_k \neq 0$. However, since

$m_1 m_2 \cdots m_k x_1 \cdots x_{k-1} = 0$, using Lemma 1.1.2 we would have the contradiction $m_1 x_1 m_2 x_2 \cdots m_{k-1} x_{k-1} m_k = 0$. Thus $0 \notin A$.

Let Q be an ideal of R maximal with respect to exclusion of A. The usual argument shows Q to be a prime ideal. Moreover, $Q \subset P$. Since P is minimal, we have $Q = P$, and so $R/P$ is a domain.

Since R has no nilpotent elements, the intersection of all prime ideals of R is 0. Hence the intersection $\bigcap P = 0$, where P runs over the minimal prime ideals. Since R is the subdirect product of these $R/P$'s and each $R/P$ is a domain, the proof of the theorem is now complete.

In a non-commutative situation, rather than speaking about zero-divisors we should be more precise, and speak about left, or right, zero-divisors. We say that $a \neq 0$ is a <u>left zero-divisor</u> if $ax = 0$ for some $x \neq 0$, and is a <u>right zero-divisor</u> if $ya = 0$ for some $y \neq 0$.

DEFINITION. An element $a \neq 0$ in R is <u>regular</u> if a is neither a left nor a right zero-divisor.

Here, too, one could talk about left regular and right regular elements.

For semi-prime rings we have an ideal-theoretic analog of Lemma 1.1.1. This is

LEMMA 1.1.3. If R is semi-prime and A, B ideals of R such that $AB = 0$ then $BA = 0$.

<u>Proof.</u> Since $AB = 0$, $(BA)^2 = BABA = 0$, whence $BA = 0$.

In fact, one can say a little more. Let $r(X) = \{y \in R \mid xy = 0 \text{ all } x \in X\}$ and $\ell(X) = \{y \in R \mid yx = 0 \text{ all } x \in X\}$ be the right and left annihilators, respectively, of the subset $X$ of $R$. Then

COROLLARY 1. If $R$ is semi-prime and $A$ is an ideal of $R$ then $r(A) = \ell(A)$.

Proof. If $C = r(A)A$, then $C$ is an ideal of $R$ and $C^2 = r(A)Ar(A)A = 0$, hence $C = 0$. That is, $r(A)A = 0$, and so $r(A) \subset \ell(A)$. Similarly, $\ell(A) \subset r(A)$, hence we conclude that $r(A) = \ell(A)$.

COROLLARY 2. If $R$ is semi-prime and $A$ is an ideal of $R$, then $A \cap r(A) = 0$.

Proof. $A \cap r(A)$ is an ideal of $R$, and $(A \cap r(A))^2 \subset Ar(A) = 0$. Therefore $A \cap r(A) = 0$.

If $T$ is any ring we shall denote the center of $T$ by $Z(T)$. We now show that certain subsets must lie in the center of a semi-prime ring.

LEMMA 1.1.4. Suppose that $R$ is semi-prime and that $a \in R$ is such that $a(ax - xa) = 0$ for all $x \in R$. Then $a \in Z(R)$, the center of $R$.

Proof. If $x, r \in R$ then $a(a(xr) - (xr)a) = 0$. However $a(xr) - (xr)a = (ax - xa)r + x(ar - ra)$; thus we get $ax(ar - ra) = 0$ for all $x, r \in R$, that is, $aR(ar - ra) = 0$. But this gives $(ar - ra)R(ar - ra) = 0$. Since $R$ is semi-prime, we conclude that $ar - ra = 0$ for all $r \in R$, hence $a \in Z(R)$.

From Lemma 1.1.4 we can derive a series of results.

LEMMA 1.1.5. Let $R$ be a semi-prime ring and let $\rho$ be a right ideal of $R$. Then $Z(\rho) \subset Z(R)$.

Proof. If $a \in Z(\rho)$ and $x \in R$ then, since $ax \in \rho$, $a(ax) = (ax)a$, that is, $a(ax - xa) = 0$. By Lemma 1.1.4 we conclude that $a \in Z(R)$.

COROLLARY. Let $R$ be a semi-prime ring and let $\rho \neq 0$ be a right ideal of $R$. If $\rho$ is commutative as a ring, then $\rho \subset Z(R)$. If, in addition, $R$ is prime, then $R$ must be commutative.

Proof. Since $\rho$ is commutative, by the lemma $\rho = Z(\rho) \subset Z(R)$. If $x, y \in R$, $a \in \rho$ then $ax \in \rho$ hence $ax \in Z(R)$; thus $(ax)y = y(ax) = ayx$ since $a \in \rho \subset Z(R)$. This yields $\rho(xy - yx) = 0$. Therefore, if $R$ is prime, since $\rho \neq 0$ is annihilated by all $xy - yx$ from the right, $xy - yx = 0$. Hence $R$ is commutative.

If $X$ is a non-empty subset of $R$, then the <u>centralizer of</u> $X$ <u>in</u> $R$, $C_R(X)$, is defined by $C_R(X) = \{a \in R \mid xa = ax \text{ all } x \in X\}$. If $a \in C_R(X)$ we say that $a$ <u>centralizes</u> $X$.

LEMMA 1.1.6. Let $R$ be a prime ring, and suppose that $a \in R$ centralizes a non-zero right ideal of $R$. Then $a \in Z(R)$.

Proof. Suppose that $a$ centralizes the non-zero right ideal $\rho$ of $R$. If $x \in R$, $r \in \rho$ then $rx \in \rho$ hence $a(rx) = (rx)a$. But $ar = ra$; we thus get that $r(ax - xa) = 0$, which is to say, $\rho(ax - xa) = 0$ for all $x \in R$. Since $R$ is prime and $\rho \neq 0$, we conclude that $ax = xa$ for all $x \in R$, hence $a \in Z(R)$.

We now show that in any ring, the annihilator of a large set of commutators is an ideal of that ring.

LEMMA 1.1.7. Let $R$ be any ring and let $u \in R$. If $V = \{a \in R \mid a(ux - xu) = 0, \text{ all } x \in R\}$, then $V$ is an ideal of $R$.

Proof. That $V$ is a left ideal of $R$ is clear. We claim that it is also a right ideal of $R$. For, let $a \in V$, $x, r \in R$; then, since $a \in V$, $a\{u(rx) - (rx)u\} = 0$. But, by the Jacobi identity for commutators, $u(rx) - (rx)u = (ur - ru)x + r(ux - xu)$. Hence $0 = a\{u(rx) - (rx)u\} = a(ur - ru)x + ar(ux - xu)$, that is, $ar(ux - xu) = 0$. Hence $ar \in V$, and $V$ is a right ideal of $R$. So, $V$ is an ideal of $R$.

As an immediate consequence we have the

COROLLARY. Let $R$ be a prime ring and suppose that $a \neq 0$ in $R$ satisfies $a(ux - xu) = 0$ for all $x \in R$. Then $u \in Z(R)$.

Proof. By Lemma 1.1.7, $V = \{r \in R \mid r(ux - xu) = 0, \text{ all } x \in R\}$ is an ideal of $R$. Since $a \in V$ and $a \neq 0$, $V \neq 0$. Since $R$ is prime and all $ux - xu \in r(V)$, we must have $ux - xu = 0$ for all $x \in R$, whence $u \in Z(R)$.

The center of a ring is, after all, the set of elements commuting with all elements of the ring. For semi-prime rings we can show that centralizing a somewhat smaller part of the ring already forces membership in the center. The result we prove below is the first, and easiest, of a large class of results of this nature which can be proved. Some recent, deeper theorems in this vein can be found in [11].

LEMMA 1.1.8. Let $R$ be a semi-prime ring, and suppose that $a \in R$ centralizes all commutators $xy - yx$, $x, y \in R$. Then $a \in Z(R)$.

Proof. If $x, y \in R$, then, since $x(ya) - (ya)x$ is a commutator, $a$ must commute with $x(ya) - (ya)x$. But $x(ya) - (ya)x = (xy - yx)a + y(xa - ax)$. By hypothesis, $a$ commutes with the left-side and the first term of the right-side of this last relation. The net result is that $a$ must commute with $y(xa - ax)$ for all $x, y \in R$. This gives us that

$(ya - ay)(xa - ax) = 0$ for all $x, y \in R$. If $V = \{r \in R \mid r(xa - ax) = 0$, all $x \in R\}$, by Lemma 1.1.7 $V$ is an ideal of $R$, and by the above, $ya - ay \in V$ for all $y \in R$. On the other hand, from the definition of $V$, all $ya - ay$ are in $r(V)$, hence in $V \cap r(V)$. But, because $R$ is semi-prime, $V \cap r(V) = 0$; hence $ay - ya = 0$ for all $y \in R$, and so $a \in Z(R)$.

Recall that a ring $R$ is said to be n-torsion free, where $n \neq 0$ is an integer, if whenever $nx = 0$, with $x \in R$, then $x = 0$.

At a small expense, namely forcing the ring to be 2-torsion free, we can obtain a considerable and important sharpening of this last lemma. The result is extremely useful.

LEMMA 1.1.9. Let $R$ be a 2-torsion free semi-prime ring. If $a \in R$ commutes with all its own commutators $ax - xa$ then $a \in Z(R)$.

Proof. Define $d: R \to R$ by $d(x) = xa - ax$ for all $x \in R$. The function $d$ so defined is a derivation of $R$, that is, $d(xy) = d(x)y + xd(y)$ for all $x, y \in R$. Our hypothesis on $a$ becomes: $d^2(x) = 0$ for all $x \in R$.

Since $d$ is a derivation on R, by Leibniz' rule, $d^2(xy) = d^2(x)y + 2d(x)d(y) + xd^2(y)$. But $d^2(xy) = d^2(x) = d^2(y) = 0$ by hypothesis. The upshot is that $2d(x)d(y) = 0$ for all $x, y \in R$. Because $R$ is 2-torsion free, we get that $d(x)d(y) = 0$ for all $x, y \in R$. Let $y = rx$ where $r$ is arbitrary in $R$; thus $d(y) = rd(x) + d(r)x$. Therefore $0 = d(x)d(rx) = d(x)\{rd(x) + d(r)x\} = d(x)rd(x)$, since $d(x)d(r) = 0$. Hence $d(x)Rd(x) = 0$. Because $R$ is semi-prime, we conclude that $d(x) = 0$ for all $x \in R$. Recalling that $d(x) = xa - ax$, we see that $a$ must in $Z(R)$.

In case of 2-torsion the result is definitely false. If $a = \begin{pmatrix} 0 & 1 \\ 1 & 0 \end{pmatrix}$ is in $F_2$, the $2 \times 2$ matrices over a field F of characteristic 2, then a commutes with all $ax - xa$, yet $a \notin Z$.

Even when there is 2-torsion we can say, in the lemma, that $2a \in Z$ but, in general, this is not too useful.

One could wonder if Lemma 1.1.9 has a natural extension to the case where $d^n = 0$, where $d(x) = xa - ax$. A little care is needed here. If $a^m = 0$ then it is trivial that $d^{2m} = 0$. In fact we can change a by adding a center element; this does not change d. For simple rings of characteristic large enough, namely, zero or larger than m, it has been shown in [3] that if $d^m = 0$ where $d(x) = xa - ax$, then there exists a $\lambda \in Z(R)$ such that $(a - \lambda)^k = 0$ where $k = [\frac{m+1}{2}]$.

## 2. Primitive Rings with Minimal One-Sided Ideals

We have as our main objective of this section, the precise description of those primitive rings which have minimal one-sided ideals. From this we shall quickly obtain the structure of such primitive rings when they are endowed with an involution.

We begin with a very well known result

LEMMA 1.2.1. Let R be a semi-prime ring. If $\rho \neq 0$ is a minimal right ideal, then $\rho = eR$ where $e^2 = e \neq 0$, and eRe is a division ring. Moreover, Re is a minimal left ideal. Conversely, if $e^2 = e \neq 0$ is such that eRe is a division ring then $\rho = eR$ is a minimal right ideal of R and $\lambda = Re$ is a minimal left ideal of R.

Proof. Since $R$ has no nilpotent ideals, $\rho^2 \neq 0$, hence $a\rho \neq 0$ for some $a \in \rho$. Since $a\rho \subset \rho$ is a right ideal of $R$, by the minimality of $\rho$ we must have $a\rho = \rho$. Thus, for some $e \in \rho$, $ae = a$, from which we deduce that $a(e^2 - e) = 0$. But $\rho_0 = \{x \in \rho \mid ax = 0\}$ is a right ideal of $R$ lying in $\rho$, and since $a\rho = \rho \neq 0$, $\rho_0 \neq \rho$. Thus $\rho_0 = 0$. Since $e^2 - e \in \rho_0$ we end up with $e^2 = e \neq 0$ (since $0 \neq a = ae$). Thus $0 \neq eR \subset \rho$ is a right ideal of $R$, so we have $\rho = eR$.

Let $\Delta = eRe$. We claim that $\Delta$ is a division ring. If $eae \neq 0 \in \Delta$ then $eaeR \neq 0$ is contained in $\rho$, hence $eaeR = \rho = eR$, and so $eaeRe = eRe$. Thus $eaeebe = e$ for some appropriate $b \in R$. This shows that $\Delta$ is a division ring.

Let $\lambda = Re$ and suppose that $0 \neq \lambda_0 \subset \lambda$ is a left ideal of $R$. Since $\lambda_0^2 \neq 0$, there is an $a \in \lambda_0$ such that $\lambda_0 a \neq 0$; but then $ea \neq 0$. Since $a \in Re$, $ae = a$, therefore $0 \neq ea = eae$. Because $eRe$ is a division ring, there is a $b \in R$ such that $ebeeae = e$; since $eae \in \lambda_0$, we get that $e \in \lambda_0$, hence $\lambda = Re \subset \lambda_0 \subset \lambda$. This shows that $\lambda = Re$ is a minimal left ideal of $R$.

The argument of the paragraph above shows that if $eRe$ is a division ring then $Re$ is a minimal left ideal of $R$. The same argument reveals that $eR$ is a minimal right ideal of $R$.

Simple artinian rings, by the very definition of artinian, have minimal right ideals. One of the first theorems one proves about such rings is that they must have a unit element. The converse is easily proved, namely, that if a simple ring with unit element has a minimal ideal then it is artinian. We do this now.

LEMMA 1.2.2. Let $R$ be a simple ring with unit element. Suppose that $R$ has a minimal right ideal. Then $R$ is an artinian ring.

<u>Proof.</u> Let $0 \neq \rho$ be a minimal right ideal of $R$. If $x \in R$ then it is immediate that $x\rho = 0$ or $x\rho$ is again a minimal right ideal of $R$. Since $R\rho \neq 0$ is an ideal of $R$, $R\rho = R$. Thus $R$ is the sum of minimal right ideals $\rho_i$ where $\rho_i = x_i\rho$ for some $x_i \in R$. Since $1 \in R$, $1 \in \rho_1 + \ldots + \rho_n$ for some $n$; this yields that $R = \rho_1 + \ldots + \rho_n$. So $R$ is the sum of a finite number of minimal right ideals, each of which is an irreducible right $R$-module. Thus $R$, as an $R$-module, has a composition series. This proves that $R$ is artinian.

Actually, as will be seen, Lemma 1.2.2 is a trivial consequence of Theorem 1.2.1. However, the proof above is so easy and elementary that it is worthwhile doing the result as we did it.

DEFINITION. Let $D$ be a division ring and suppose that $V$ is a left vector space over $D$ and $W$ a right vector space over $D$. We say that the function $( \, , \, ): V \times W \to D$ is an <u>inner product</u> between $V$ and $W$ if it is additive in each component, and if

1. $(\alpha v, w) = \alpha(v, w)$
2. $(v, w\beta) = (v, w)\beta$

for all $v \in V$, $w \in W$ and $\alpha, \beta \in D$.

DEFINITION. $V$ and $W$ as above are said to be <u>dual</u> if $(v, W) = 0$ forces $v = 0$ and $(V, w) = 0$ forces $w = 0$.

Suppose that $V$ and $W$ are dual spaces. Let $A(V) = \mathrm{Hom}_D(V, V)$ act on $V$ on the right and $A(W) = \mathrm{Hom}_D(W, W)$ act on $W$ on the left.

If $a \in A(V)$, then the element $a^* \in A(W)$ is called an <u>adjoint</u> of $a$ if $(va, w) = (v, a^*w)$ for all $v \in V$ and $w \in W$.

While elements in $A(V)$ may fail to have adjoints in $A(W)$, by the dual nature of $V$ and $W$ if $a \in A(V)$ does have an adjoint in $A(W)$, that adjoint must be unique. For, if $(va, w) = (v, a^*w) = (v, tw)$ for all $v \in V$, $w \in W$, with $a^*, t$ in $A(W)$, then $(v, (a^* - t)w) = 0$. Hence $a^* = t$ follows.

In what follows, $V$ and $W$ are given dual spaces

DEFINITION. $a \in A(V)$ is said to be <u>continuous</u> if it has an adjoint in $A(W)$.

Note that the set of continuous linear transformations on $V$ form a subring of $A(V)$. Call this subring $L_W(V)$. We could identify $L_W(V)$ as the set of linear transformations continuous with regard to a suitable topology on $V$, however we shall have no need of such a description in what is to follow.

DEFINITION. An element $a \in A(V)$ is said to be of <u>finite range</u> if $\dim_D(Va) < \infty$.

Let $S_W(V) = \{a \in L_W(V) = a$ is of finite range$\}$.

We come to our first structure theorem. The theorem is due to Jacobson [17, 18]; the proof we give is one by Kaplansky [19].

THEOREM 1.2.1. Let $V$ and $W$ be dual spaces over a division ring $D$ and let $R$ be a subring of $A(V)$ such that $S_W(V) \subset R \subset L_W(V)$. Then $R$ is a primitive ring containing a minimal left ideal. Moreover, $S_W(V)$ is the unique minimal ideal of $R$.

Conversely, given a primitive ring R with minimal left ideal, then we can find a division ring D and dual spaces V and W over D such that $S_W(V) \subset R \subset L_W(V)$.

Proof. Suppose that $S_W(V) \subset R \subset L_W(V)$. If $v_0 \neq 0$ is in V, and $w_0 \neq 0$ is in W then the mapping $a: V \to V$ defined on all $v \in V$ by $va = (v, w_0)v_0$ is of 1-dimensional range. It is immediate to verify that a has an adjoint $a^*: W \to W$ defined by $a^*w = w_0(w, v_0)$. Thus $a \in S_W(V) \subset R$. Given $v_1 \neq 0$ and $v_0 \neq 0$ in V, since $(v_1, W) \neq 0$ there is a $w_0 \in W$ with $(v_1, w_0) = 1$. But then the map $a: V \to V$ defined by $va = (v, w_0)v_0$ takes $v_1$ onto $v_0$. Hence R acts transitive on V. Since $R \subset L_W(V) \subset A(V)$, R acts faithfully on V. Thus V is an irreducible and faithful right R-module, hence R is right primitive.

If $b \in S_W(V)$ is of 1-dimensional range then $vb = \lambda(v)v_0$ where $\lambda(v) \in D$, from every $v \in V$. Hence, for $w \in W$, $\lambda(v)(v_0, w) = (vb, w) = (v, b^*w)$. Pick $w_0 \in W$ with $(v_0, w_0) = 1$, and let $w_1 = b^*w_0$. Thus $\lambda(v) = (v, w_1)$ by the above, so that $vb = (v, w_1)v_0$; that is, b is of the form described above.

Let $v_0 \neq 0$ be in V and let $\lambda = \{a \in R \mid Va \subset Dv_0\}$. Since $R \supset S_W(V)$, $\lambda \neq 0$. Trivially, $\lambda$ is a left ideal of R. We claim that $\lambda$ is a minimal left ideal of R. For, if $a \neq 0$ and $b \neq 0$ are in $\lambda$, by the preceding paragraph $va = (v, w_0)v_0$ and $vb = (v, w_1)v_0$ for suitable $w_0, w_1 \in W$, for all $v \in V$. Pick $v' \in V$ such that $(v', w_0) = 1$ and let $t: V \to V$ be defined by $vt = (v, w_1)v'$. Thus $t \in S_W(V) \subset R$ and $vta = ((v, w_1)v', w_0)v_0 = (v, w_1)(v', w_0)v_0 = (v, w_1)v_0 = vb$. Hence $ta = b$. This shows that $\lambda$ is a minimal left ideal of R. Since R is right primitive and has a minimal left ideal, R is also left primitive.

Let $A \neq 0$ be an ideal of $R$. Since $R$ is primitive, $AS_W(V) \neq 0$, and since $AS_W(V) \subset A \cap S_W(V)$, we have that $A$ has elements of finite range, and so has elements of 1-dimensional range. To show that $A \supset S_W(V)$ it is enough to show that $A$ has all elements in $R$ of 1-dimensional range.

Let $0 \neq a \in A$ be of 1-dimensional range; as we saw, $va = (v, w_0)v_0$ for some $0 \neq v_0 \in V$, $0 \neq w_0 \in W$. If $b \in S_W(V)$ is of 1-dimensional range then $vb = (v, w_1)v_1$ for suitable $v_1 \in V$, $w_1 \in W$. Since $R$ is transitive on $V$, $v_0c = v_1$ for some $c \in R$, hence $vac = (v, w_0)v_0c = (v, w_0)v_1$. As is easy to see, there is a $t \in R$ such that $t^*w_0 = w_1$. Hence $vtac = (vt, w_0)v_1 = (v, t^*w_0)v_1 = (v, w_1)v_1 = vb$. Hence $b = tac$ so is in $A$ since $a \in A$ and $A$ is an ideal of $R$. Thus $A \supset S_W(V)$ and so $S_W(V)$ is the unique minimal ideal of $R$.

In the other direction, suppose that $R$ is a left primitive ring with minimal left ideal $\lambda$. Since $R$ is primitive ring, it has no nilpotent ideals, hence $\lambda = Re$ where $e^2 = e$ and where $D = eRe$ is a division ring. Let $V = eR$, $W = Re$; $V$ is a left vector space over $D$ and $W$ a right vector space over $D$, by Lemma 1.2.1. Define, for $v = ea$, $w = be$, $(v, w) = (ea, be) = eabe \in D$. This is trivially an inner product. Since $R$ is primitive, $R$ is prime, so if $(v, W) = 0$ and $v = ea$, then $eaRe = 0$, whence $ea = 0$, that is, $v = 0$. Similarly $(V, w) = 0$ forces $w = 0$. Thus $V$ and $W$ are dual. If $x \in R$ then $x$ induces a D-linear transformation $T_x$ on $V$ by $(ea)T_x = eax$. Given $v = ea$, $w = be$ then $(vT_x, w) = (eax, be) = eaxbe = (ea, xbe) = (v, T_x^* w)$ whence $T_x^* w = xw$. Thus $\{T_x \mid x \in R\} \subset L_W(V)$. Since $R$ is primitive $R$ is isomorphic to $\{T_x \mid x \in R\}$, hence $R \subset L_W(V)$.

To see that $R \supset S_W(V)$ we only need to see that every $a \in L_W(V)$ of 1-dimensional range is in R. If $va = Dv_0$ and $a \in L_W(V)$ then $va = (v, w_0)v_0$, that is, $era = (erb_0e)ec_0$ where $w_0 = b_0e$, $v_0 = ec_0$. Thus $era = er(b_0ec_0)$; this gives that $a = T_{b_0ec_0}$, so is in R.

The theorem is now completely proved.

The theorem has several interesting implications which we list as corollaries.

COROLLARY 1. If R is a simple ring with minimal right ideal then $R = S_W(V)$ where $V = eR$, $W = Re$ are minimal right and left ideals of R.

Note that Lemma 1.2.2 is an immediate consequence of this corollary. For, if R is simple with unit element and has a minimal right ideal then, by this corollary, $R = S_W(V)$. Hence $1 \in R$ must be of finite-dimensional range, that is, $V = V1$ is finite-dimensional. But then R is artinian.

The consequence of Theorem 1.2.1 which we shall use most often is

COROLLARY 2. Let R be a primitive ring with minimal right ideal. Then either R is isomorphic to $D_n$, where D is a division ring, for some n, or, for any $m > 0$, R contains a subring isomorphic to $D_m$.

<u>Proof.</u> Let V and W be dual vector spaces such that $S_W(V) \subset R \subset L_W(V)$. If V is finite-dimensional over D, then we get $R = S_W(V) = L_W(V) = A(V)$, the ring of all linear transformations on V over D. Hence $R \approx D_n$.

On the other hand, if $\dim_D V = \infty$, let $v_1, \ldots, v_m \in V$ be linearly independent over $D$ and let $V_m$ be the subspace they span over $D$. Let $V = V_m \oplus U$ and let $A_m = \{a \in R \mid Ua = 0\}$. By the theorem we then get, since $R \supset S_W(V)$, that $A_m \approx D_m$.

DEFINITION. If $R$ is any ring, a mapping $*: R \to R$ is called an <u>involution</u> if :

1. $a^{**} = a$
2. $(a+b)^* = a^* + b^*$
3. $(ab)^* = b^* a^*$

for all $a, b \in R$.

Let $D$ be a division ring having an involution $*$ and let $V$ be a left-vector space over $D$. Suppose that $V$ is self-dual, in that there is a non-degenerate inner product defined on $V$. We say that this inner product is <u>Hermitian</u> if $(v, w) = (w, v)^*$ for $v, w \in V$. We say this inner product is <u>alternate</u> if $(v, w) = -(w, v)$ for all $v, w \in V$.

If the form on $V$ is alternate, it is easy to show that $D$ must, in fact, be a field and $a^* = a$ for every $a \in D$.

The next theorem is due to Kaplansky; the first part of the proof comes from an argument due to Rickart.

THEOREM 1.2.2. Let $R$ be a primitive ring with involution $*$, and suppose that $R$ has a minimal right ideal. Then there exists a vector space $V$ over a division ring $D$ which is self-dual with respect to a Hermitian or alternate form in such a way that:

1. $S_V(V) \subset R \subset L_V(V)$
2. the $*$ of $R$ is the adjoint (on R) relative to this form.

Furthermore, the Hermitian and alternate cases are mutually exclusive, occuring according as there is, or is not, a primitive (minimal) idempotent e such that $e^* = e$.

Proof. If there exists a self-adjoint idempotent $e^* = e = e^2$ such that $V = eR$ is a minimal right ideal, then $D = eRe$ is a division ring, and V is a left-vector space over D. Moreover, $D^* = (eRe)^* = e^*Re^* = eRe = D$, hence the * of R induces an involution on D. We define an inner product on V by means of: if $v = ex$, $w = ey$ are in V then $(v, w) = exy^*e = vw^*$. That this is an inner product is trivial from the defining properties of an involution. It is Hermitian, for $(v, w) = vw^* = (wv^*)^* = (w, v)^*$. If $r \in R$, $(vr, w) = vrw^* = v(w^*r^*)^* = (v, wr^*)$, so the * on R coincides with the adjoint corresponding to the form. Finally, since R is primitive, hence prime, the form is non-degenerate, for if $(V, w) = 0$ then $Vw^* = 0$, that is, $eRw^* = 0$; this forces $w^* = 0$, and so $w = 0$. In this case, by a direct application of Theorem 1.2.1 to our situation, we have Theorem 1.2.2.

We now claim that if any minimal right ideal $\rho$ contains a self-adjoint element a such that $a\rho \neq 0$ then $\rho = eR$ where $e^2 = e = e^*$ (and so we are back in the situation discussed above). Suppose then that $a^* = a$ is in $\rho$ and $a\rho \neq 0$. By the minimality of $\rho$ we must have $a\rho = \rho$, whence $at = a$ for some $t \in \rho$ where $t^2 = t$. Thus $tR = \rho$ and t is a left unit for $\rho$; therefore $ta = a$. But then $a = a^* = (ta)^* = a^*t^* = at^* = (at)t^*$. From this, $e = tt^* \neq 0$ is a self-adjoint idempotent in $\rho$.

Thus we may assume that if $\rho$ is a minimal right ideal of R and if $a^* = a \in \rho$, then $a\rho = 0$. We claim that $xx^* = 0$ for all $x \in \rho$. Since $xx^* \in \rho$ and is self-adjoint, $xx^*\rho = 0$, hence $xx^*x = 0$; thus

$x^*xx^* = (xx^*x)^* = 0$. Now, if $xx^* \neq 0$ then $xx^*R = \rho$ by the minimality of $\rho$; this yields $x^*\rho = x^*xx^*R = 0$, in other words $x^*\rho = 0$ for all $x \in \rho$.

Let $0 \neq x \in \rho$; then, since $\rho = xR$ is a minimal right ideal of R, $\rho^* = (xR)^* = Rx^*$ is a minimal left ideal of R. From this, via Lemma 1.2.1, it is easy to see that $\rho = x^*R$ is a minimal right ideal of R. If $y \in \rho_1$ then, as above, $y^*\rho_1 = 0$. Let $y = x^* \in \rho_1$; thus $(x^*)^*x^*R = (x^*)^*\rho_1 = 0$, hence $xx^*R = 0$ and so $xx^* = 0$. So we have seen that $xx^* = 0$ for all $x \in \rho$.

If $\rho = eR$ is a minimal right ideal of R, by the primitivity of R, there is a $b \in R$ such that $ebe^* \neq 0$. If $x, y \in R$ then there is a unique element $eze$ such that $exy^*e = ezebe^*$. This follows easily from the fact that $eRe$ is a division ring. Define the map $(\ ,\ )$ on $V \times V$ into $D = eRe$ by $(x, y) = eze$, for $x, y \in V$, where $exy^*e^* = ezebe^*$. It is trivial to verify that this defines an inner product on V. Since R is primitive, this inner product is non-degenerate. Since $(ex, ex) = 0$ — from $exx^*e^* = (ex)(ex)^* = 0$ — the form is alternate. Applying Theorem 1.2.1 we have completely proved Theorem 1.2.2.

We have, in this context, an analog for Corollary 2 to Theorem 1.2.1.

COROLLARY. If R is primitive with involution * and has a minimal right ideal then, either R is isomorphic to $D_m$ for some m, where D is a division ring or, for every $n > 0$, R has a subring $A(n)$, invariant re *, such that

1. $A(n) \approx D_n$, if $e^* = e$ for a primitive idempotent e in R, with * of transpose type, or

2. $A(n) \approx D_{2n}$, D a field, if $e^* \neq e$ for every primitive idempotent e in R, with * symplectic.

### 3. Generalized Polynomial Identities

In the preceding section we obtained a complete description of primitive rings having 1-sided minimal ideals, and of such rings with involution. In order to be able to exploit this characterization we need some method either of recognizing such primitive rings or of making them arise. A very fruitful such means comes via a theorem of Martindale [22], which generalizes an earlier result due to Amitsur [1]. This result of Martindale has a wide scope for application and lends itself readily to a rather large variety of problems. We shall find it useful in the type of studies of rings with involution that we shall make.

Before getting down to the problem of generalized polynomial identities, we must develop some material on a general "ring of quotients" for prime rings. This is related to work of Utumi [35] and his ring of quotients. We shall work in the context of prime rings; however these notions can be developed more generally, as Amitsur has shown [2].

Let R be a prime ring and let $\mathcal{M}$ be the set of all pairs $(U, f)$ where $U \neq 0$ is an ideal of R and $f: U \to R$ is a right R-module map of U into R.

We define an equivalence relation on $\mathcal{M}$ by declaring $(U,f) \sim (V,g)$, for $(U, f), (V, g)$ in $\mathcal{M}$, if $f = g$ on some ideal $W \neq 0$ of R where

$W \subset U \cap V$. Since R is prime it is trivial to verify that this defines an equivalence relation on $\mathfrak{M}$.

Let Q be the set of equivalence classes of $\mathfrak{M}$. We denote the equivalence class of (U, f) as $\tilde{f}$. We now propose to make of Q a ring, in fact a ring containing R. To do so we must define an addition and multiplication in Q.

If $\tilde{f} = cl(U, f)$, $\tilde{g} = cl(V, g)$ are in Q define $\tilde{f} + \tilde{g} = cl(U \cap V, f+g)$ and $\tilde{f} \cdot \tilde{g} = cl(VU, fg)$. It is straight-forward, making use of the primeness of R, to show that Q is an associative ring with unit element relative to these operations.

If $a \in R$ define $\lambda_a : R \to R$ by $\lambda_a(x) = ax$. Clearly $\lambda_a$ is a right R-module map of R into itself. Let $\tilde{a} = cl(R, \lambda_a)$. If $\tilde{a} = 0$ then, by the definition of our equivalence, $aU = 0$ for some ideal $U \neq 0$ of R. By the primeness of R we conclude that $a = 0$. Thus the mapping $R \to Q$ given by $a \to \tilde{a}$ is 1-1. Since, clearly, $\widetilde{ab} = \tilde{a}\tilde{b}$, we have that R is embedded isomorphically in Q. We consider $R \subset Q$.

A very important property that is enjoyed by R relative to Q in this embedding, and one which falls from the very construction of Q, is:

<u>if</u> $q \neq 0 \in Q$ <u>then there exists an ideal</u> $U \neq 0$ <u>of</u> R <u>such that</u> $0 \neq qU \subset R$.

In fact, since R is prime, if we have $q_1, \ldots, q_n$ in Q there is an ideal $U \neq 0$ of R such that each $0 \neq q_i U \subset R$.

We claim that Q is prime. For, if $q_1 Q q_2 = 0$, $q_1, q_2 \in Q$ then, if $q_1 \neq 0$, $q_2 \neq 0$, there is an ideal U of R such that $0 \neq q_1 U \subset R$ and $0 \neq q_2 U \subset R$. Hence $(q_1 U)R(q_2 U) \subset q_1 Q q_2 U = 0$. This contradicts the primeness of R.

Let $C = Z(Q)$; $C$ consists, clearly, of all pairs $[U, f]$ where $f$ is an $R$-bi-module map of $U$ into $R$. Since $Q$ is prime $C$ must be an integral domain. In fact, a non-zero element of $C$ cannot be a zero divisor in $Q$. Also, $1 \in C$. However, much more can be said. This is

LEMMA 1.3.1. $C$ is a field.

Proof. Let $c \neq 0$ be in $C$; thus there exists an ideal $U$ of $R$ such that $0 \neq cU \subset R$. However, since $c$ commutes with all elements of $Q$, and so, with all elements of $R$, $V = cU \neq 0$ is an ideal of $R$. Consider the map $f: V \to R$ defined by $f: cu \to u$. Clearly $f$ is a right $R$-module map of $V$ into $R$. Let $d = cl(V, f)$. By the definition of $d$, $dc = 1$ since $f(cu) = u$, so $(dc - 1)U = 0$.

DEFINITION. $C$ is the extended centroid of $R$ and $S = RC$ is the central closure of $R$.

It is immediate that $S$ is prime. If $R$ has a unit element, then $C = Z(S)$. If $R$ is a simple ring with unit element, since the only non-zero ideal of $R$ is $R$ itself, we quickly can verify that $Q = S = R$, that is, $R$ is its own central closure.

LEMMA 1.3.2. Suppose that $a_i, b_i$ are non-zero elements in $R$ such that $\sum a_i x b_i = 0$ for all $x \in R$. Then the $a_i$'s are linearly dependent over $C$, and the $b_i$'s are linearly dependent over $C$.

Proof. We show that the $a_i$'s are linearly dependent over $C$. If not, there is a minimal $n$ and elements $a_1, \ldots, a_n \in R$ linearly independent over $C$ such that $\sum_1^n a_i x b_i = 0$ for all $x \in R$, where the $b_i$ are non-zero elements of $R$. Since $R$ is prime, $n > 1$.

Suppose that $x_j, y_j \in R$ are such that $\sum x_j b_1 y_j = 0$. If $r \in R$ then

$$0 = \sum a_1 r x_j b_1 y_j = -\sum_{i=2}^{n} a_i r(\sum x_j b_i y_j)$$

since $\sum_{i=1}^{n} a_i r x_j b_i = 0$. Since we have a shorter relation than $n$, we have that $\sum_j x_j b_i y_j = 0$ for all $i$. Hence the map $\gamma_i : Rb_1R \to R$ defined by $\gamma_i(\sum u_j b_1 v_j) = \sum u_j b_i y_j$ is well-defined. It is trivial that $\gamma_i$ is a module map of the ideal $Rb_1R$ into $R$, hence $\gamma_i$ gives us an element — which we also denote by $\gamma_i$ — in $Q$. It is trivial to verify that $\gamma_i$ is in fact in $C$. Moreover, by its definition, $\gamma_i b_1 = b_i$. Thus $0 = \sum a_i x b_i = \sum a_i x \gamma_i b_1 = (\sum \gamma_i a_i) x b_1$. By the primeness of $S$ we get that $\sum \gamma_i a_i = 0$; since the $a_i$ are linearly independent over $C$, we must have $\gamma_i = 0$. But then, by the definition of $\gamma_i$, $Rb_iR = 0$, giving us the contradiction $b_i = 0$. This proves the lemma.

A very special case of the lemma, but one which occurs often, is the

COROLLARY. If $a, b \in R$ are such that $axb = bxa$ for all $x \in R$, and $a \neq 0$, then $b = \lambda a$ for some $\lambda \in C$.

The next theorem looks extremely special — in fact, it is. But, strangely enough, it will keep recurring in the next chapter. The result we get is a special case of the more general result due to Martindale which is expressed in Theorem 1.3.2. However, we could get by with it, rather than with the full Martindale theorem, in our situation. For the sake of general interest — Martindale's theorem is very important — we shall also derive the result of Martindale.

THEOREM 1.3.1. Suppose that $R$ is a prime ring and that $a \neq 0$ in $R$ is such that $axaya = ayaxa$ for all $x, y \in R$. Then $S = RC$ is a primitive ring with minimal right ideal, and the commuting ring of $S$ on this right ideal is merely $C$ itself.

Proof. Fixing $x$ in the relation $(axa)ya - ay(axa) = 0$, applying the corollary to Lemma 1.3.2, we obtain that $axa = \lambda(x)a$ where $\lambda(x) \in C$, for every $x \in R$. Since $S = RC$, we also have $aya = \lambda(y)a$ for every $y \in S$, that is, $aSa \subset Ca$. Since $S$ is prime, $ay_0a \neq 0$ for some $y_0 \in S$; thus $ay_0a = \lambda a$ where $\lambda \neq 0$. If $x_0 = \lambda^{-1}y_0$, then $ax_0a = a$, hence $e = ax_0$ is an idempotent. Moreover, $eSe = ax_0Sax_0 \subset Cax_0 = Ce$. Thus, by Lemma 1.3.1, $eS$ is a minimal right ideal of $S$, and $Ce$ is the commuting ring of $S$ on $eS$. Because $S$ is prime and has a minimal right ideal, $S$ is primitive. This finishes the proof of the theorem.

A special case of this theorem, which is of independent interest is the

COROLLARY. Let $R$ be a simple ring with unit and suppose that for some $a \neq 0$ in $R$ we have $axaya = ayaxa$ for all $x, y \in R$. Then $R$ is isomorphic to the ring of all $n \times n$ matrices over a field.

Proof. As we noted earlier, $R = S$ since $R$ is simple and has a unit element. By the theorem, $R$ has a minimal right ideal; thus, by Lemma 1.2.2, $R$ is simple artinian. Also, by the theorem, the commuting ring of $R$ on an irreducible module is $C = Z(R)$ itself. Thus, by Wedderburn's theorem, $R$ is the ring of all $n \times n$ matrices over the field $C$.

We now turn in the direction of the more general result due to Martindale. The crucial step is a result which, in a certain sense, is a direct generalization of Lemma 1.3.2.

LEMMA 1.3.3. Let $a_1, \ldots, a_n \in S = RC$ be linearly independent over $C$, and suppose that $b_1, \ldots, b_n \in S$, with $b_1 \neq 0$. Suppose that $V = \{\sum_1^n a_i x b_i \mid x \in S\}$ is finite-dimensional over $C$. Then:

1. $V \neq 0$

2. $S$ has s minimal right ideal $eS$, where $e^2 = e \neq 0$ (and so, $S$ is primitive)

3. $eSe$ is a division algebra finite-dimensional over $C$.

Proof. We proceed by induction on $n$. If $n = 1$ then $B = a_1 S b_1$ is finite-dimensional over $C$. Since $S$ is prime and $a = a_1 \neq 0$, $b = b_1 \neq 0$, we have that $B \neq 0$. Since $S$ is prime there exists a $c \in S$ such that $bca \neq 0$. Then $E = Bc = aSbc$ is a ring, $E^2 \neq 0$ and, by the finite-dimensionality of $B$, $E$ is finite-dimensional over $C$. If $N$ is the radical of $E$, then $N$ is nilpotent and $E/N \neq 0$ is a finite-dimensional, semi-simple algebra over $C$. Thus $E/N$ has an idempotent — in fact, a unit element — which we can pull back to an idempotent $f$ in $E$. Since $f \in E = aSbc$, $fSf \subset aSbc$, so $fSf$ is finite-dimensional over $C$. But $fSf$ is prime, hence is simple; thus, by Wedderburn's theorem, $fSf \approx \Delta_m$, where $\Delta$ is a division algebra finite-dimensional over $C$. Thus $fSf$ has a minimal idempotent $e$. Now $efSfe \approx \Delta e$, so is isomorphic to $\Delta$. Since $e \in fSf$, $ef = fe = e$, hence $efSfe = eSe$. In short, $eSe$ is the division ring $\Delta e$, so is finite-dimensional over $C$. By Lemma 1.2.1, $eS$ is a minimal right ideal of $S$. Thus, for $n = 1$, the lemma is proved.

Suppose, then, that $n > 1$. Let $v_1, \ldots, v_k$ be a basis over $C$ of $V = \{\sum a_i x b_i \mid x \in S\}$. Thus, for $x \in S$,

$$(1) \qquad \sum_1^n a_i x b_i = \sum_1^k \lambda_j(x) v_j \ , \quad \lambda_j(x) \in C.$$

Pick $y \in S$, and multiply (1) onthe fight by $yb_1$; we get

$$(2) \qquad \sum_1^n a_i x b_i y b_1 = \sum_1^k \lambda_j(x) v_j y b_1 \ .$$

However,

$$(3) \qquad \sum_1^n a_i (x b_1 y) b_i = \sum \lambda_j (x b_1 y) v_j \ .$$

Subtracting (3) from (2) yields

$$(4) \qquad \sum_{i=2}^n a_i x (b_i y b_1 - b_1 y b_i) \in V + V y b_1 \ ,$$

hence $\{\sum_{i=2}^n a_i x (b_i y b_1 - b_1 y b_i) \mid x \in S\}$ is finite-dimensional over $C$.

If $b_i y b_1 = b_1 y b_i$ for all $y \in S$, and all $i = 1, 2, \ldots, n$, then by the corollary to Lemma 1.3.2, since $b_1 \neq 0$, $b_i = \mu_i b_1$ where $\mu_i \in C$. Thus

$$\sum_1^n a_i x b_i = \sum_1^n a_i x \mu_i b_1 = \{\sum (\mu_i a_i)\} x b_1 \quad ;$$

since $a_1, \ldots, a_n$ are linearly independent over $C$ and $\mu_1 = 1$, $a = \sum \mu_i a_i \neq 0$. But then $\{a x b_1 \mid x \in S\}$ is finite-dimensional over $C$, and we are back to the case $n = 1$. So we may assume that $c_i = b_i y b_1 - b_1 y b_i \neq 0$ for some $y \in S$ and some $i$, say $i = 2$.

However (4) then tells us that $\{\sum_2^n a_i x c_i \mid x \in S\}$ is finite-dimensional over $C$, and $c_2 \neq 0$. Since the number of $a$'s involved is $n-1$, by induction we have proved the theorem.

Let, again, R be prime and $S = RC$ the central closure of R. Consider $S\langle x\rangle = S *_C \{x_1, x_2, \ldots, x_n, \ldots\}$, the free product over C of S and $\{x_1, \ldots, x_n, \ldots\}$. A typical element in $S\langle x\rangle$ is a sum of monomials of the form $\gamma a_{i_0} x_{j_1} a_{i_1} x_{j_2} \cdots x_{j_n} a_{i_n}$ where $\gamma \in C$, the $a_{i_k} \in S$ and $x_{j_k} \in \{x_1, \ldots, x_n\}$ (we can allow $a_{i_k}$ to be 1 formally also).

DEFINITION. We say that S satisfies a generalized polynomial identity if there exists an $f \neq 0$ in $S\langle x\rangle$ with $f(s_1, \ldots, s_n) = 0$ for all $s_i \in S$.

As for polynomial identities, we can speak about the degree of a generalized polynomial identity (henceforth, abbreviated to GPI), homogeneity and multilinearity. As we prove for polynomial identities, so can we for GPI's, that if S satisfies a GPI then it satisfies a homogeneous, multilinear one.

We now prove a lovely theorem due to Martindale [22].

THEOREM 1.3.2. Let R be a prime ring and $S = RC$ its central closure. Then S satisfies a GPI over C if and only if S is a primitive ring, containing a minimal right ideal eS, $e^2 = e$, such that eSe is a division algebra finite-dimensional over C.

Proof. One way is fairly trivial. If eS and eSe are, as above, then eSe satisfies a suitable standard identity, since it is finite-dimensional over C. If eSe satisfies $\sum_{\sigma \in S_n} (-1)^\sigma x_{\sigma(1)} \cdots x_{\sigma(n)}$ then, for $s \in S$, $\sum (-1)^\sigma es_{\sigma(1)}es_{\sigma(2)}e \cdots es_{\sigma(n)}e = 0$. Hence S satisfies the GPI $\sum (-1)^\sigma ex_{\sigma(1)}ex_{\sigma(2)} \cdots ex_{\sigma(n)}e$ over C.

We now proceed to the important, other direction of the theorem. Suppose that S satisfies a GPI of minimal degree n; as we remarked above, we may assume this GPI to be homogeneous and multilinear. Hence we may assume that we can write this minimal GPI for S over C as $f(x_1, \ldots, x_n) = \sum_1^m a_i x_1 f_i(x_2, \ldots, x_n) + g(x_1, \ldots, x_n)$ where:

1. $a_1, \ldots, a_m \in S$ are linearly independent over C
2. each $f_i$ is of degree n-1
3. no monomial in the expression for g has $x_1$ as its first variable.

We break up f even finer, writing it as:

$$(1) \qquad f(x_1, \ldots, x_n) = \sum_1^m a_i x_1 f_i + \sum_1^k g_i x_1 b_i + \sum p_i x_1 q_i$$

where

4. $b_1, \ldots, b_k \in S$ are linearly independent over C
5. $g_i$ is of degree n-1
6. $p_i$, $q_i$ are both of positive degree.

Hence, in the last sum in (1), $x_1$ is never the first nor last variable.

If $y \in S$, multiply (1) by $yb_1$ from the right; so, if $s_1, \ldots, s_n \in S$

$$(2) \qquad \sum a_i s_1 f_i(s_2, \ldots, s_n) y b_1 + \sum g_i(x_2, \ldots, s_n) s_1 b_i y b_1 + \sum p_i s_1 q_i y b_1 = 0$$

where $p_i$, $q_i$ in this last sum are evaluated at $(s_1, \ldots, s_n)$.

On the other hand, substituting $s_1b_1y$ for $x_1$ in (1), $s_2, \ldots, s_n$ for $x_2, \ldots, x_n$ respectively, gives us

$$(3) \qquad \sum_1^n a_i s_1 b_1 y f_i(s_2, \ldots, s_n) + \sum_1^k g_i(s_2, \ldots, s_n) s_1 b_1 y b_i + \sum p_i s_1 b_1 y q_i = 0 .$$

Subtract (2) from (3); we get (leaving out the variables $s_2, \ldots, s_n$ in the $f_i, g_i, p_i, q_i$)

$$(4) \qquad \sum_1^m a_i s_1 (f_i y b_1 - b_1 y f_i) + \sum_2^k g_i s_1 (b_i y b_1 - b_1 y b_i) + \sum p_i s_1 (q_i y b_1 - b_1 y q_i) = 0 .$$

If $f_1 y b_1 - b_1 y f_1 = 0$ for all $y \in S$ and all $s_2, \ldots, s_n \in S$ then, by the corollary to Lemma 1.3.2, $f_1(s_2, \ldots, s_n) = \lambda(s_2, \ldots, x_n) b_1$, $\lambda(s_2, \ldots, s_n) \in C$. Now, since $f$ is a GPI of minimal degree for $S$, and $\deg f_1 < \deg f$, $f(r_2, \ldots, r_n) \neq 0$ for some $r_2, \ldots, r_n \in S$. Let $h(x_2) = f_1(x_2, r_3, \ldots, r_n)$. Since $h(r_2) = f_1(r_2, \ldots, r_n) \neq 0$, $h \not\equiv 0$ in $S<x>$. Now the form of $h(x_2)$ is given by $h(x_2) = \sum c_i x_2 d_i$, $c_i, d_i \in S$ and the $c_i$ linearly independent over $C$. By the above, $h(s_2) = f(s_2, r_3, \ldots, r_n) = \lambda(s_2, r_3, \ldots, r_n) b_1$ so $\{\sum c_i x d_i \mid x \in S\}$ is 1-dimensional over $C$. By Lemma 1.3.3 we are done, and the theorem would be proved. So we may assume that

$f_1(s_2, \ldots, s_n) y b_1 - b_1 y f(s_2, \ldots, s_n) \neq 0$ for some $s_2, \ldots, s_n \in S$ and some $y = t_0 \in S$. Let:

$$f_i' = f_i t_0 b_1 - b_1 t_0 f_i$$

$$b_i' = b_i t_0 b_1 - b_1 t_0 b_i$$

$$q_i' = q_i t_0 b_1 - b_1 t_0 q_i .$$

By (4), $S$ satisfies the non-trivial GPI

(5) $$\sum_1^n a_i x_1 f_i' + \sum_2^k g_i x_1 b_i' + \sum p_i x_1 q_i' .$$

We know that (5) is non-trivial, for were it trivial, by the nature of equality in $S<x>$, $\sum_1^n a_i x_1 f_i'$ would be trivial. Since $f_1' \neq 0$, if $\{\sum a_i x f_i' \mid x \in S\} = 0$, we would violate the conclusion of Lemma 1.3.3. So (5) is indeed non-trivial.

Now look at (5) a moment. In going from (1) to (5), what did we accomplish? First note that the variables $x_1, \ldots, x_n$ stay in the same order in every monomial present, and $a_1, \ldots, a_n$ were unchanged. The important change was in the second sum, for in this passage, we have reduced $k$ by at least 1. Continuing in this process we come to a GPI satisfied by $S$ of the form $\sum_1^n a_i x_1 f_i(x_2, \ldots, x_n) + g(x_1, \ldots, x_n)$ where $x_1$ is <u>never the first nor the last</u> variable in any monomial of $g(x_1, \ldots, x_n)$, and the order of the variables in any monomial appearing is as they were in the original identity. Let $x_1, \ldots, x_r$ be these variables which appear as first variables of monomials, $r \leq n$. Applying the above process to each of $x_1, \ldots, x_r$ gives us a GPI for $S$ of the form:

(6) $$\sum a_i x_1 f_i + \sum b_i x_2 g_i + \ldots + \sum d_i x_r h_i$$

where:

1. $\{a_i\}, \{b_i\}, \ldots, \{d_i\}$ are each sets of elements of $S$ linearly independent over $C$.

2. $f_i, g_i, \ldots, h_i$ are homogeneous of degree $n-1$.

3. $x_1, \ldots, x_r$ are never last variables in the monomials of the $f_i, g_i, \ldots, h_i$.

Since some variable must appear last, we clearly have $r < n$. Since $f$ is a minimal GPI for $S$, and $\deg f_1 < \deg f$, there are elements $r_2, \ldots, r_n \in S$ such that $f_1(r_2, \ldots, r_n) \neq 0$. Let $f_i', g_i', \ldots, h_i'$ be obtained from $f_i, g_i, \ldots, h_i$ by letting $x_n = r_n$ and the other variables $x_2, \ldots, x_{n-1}$ free. Since $r < n$ we can do this. We claim that

$$(7) \qquad \sum a_i x_1 f_i' + \sum b_i x_2 g_i' + \ldots + \sum d_i x_n h_i'$$

is a non-trivial GPI of degree $n-1$ satisfied by $S$. If it is trivial, then $\sum a_i x_1 f_i'$ is trivial, so if $c_i = f_i'(r_2, \ldots, r_n)$ then $c_1 \neq 0$ and $\{\sum a_i x c_i \mid x \in S\} = 0$, violating the conclusion of Lemma 1.3.3. Hence (7) is non-trivial. It vanishes on $S$; this is clear from (6) and the form of $f_i', g_i', \ldots, h_i'$. Since the degree of (6) is $n-1$ we have reached a contradiction. The theorem is now proved.

Martindale's theorem has some interesting consequences. We do some of its immediate implications now. The first is a result due to Amitsur [1].

THEOREM 1.3.3. Let $R$ be a primitive ring; suppose that $R$ acts densely on the vector space $V$ over the division ring $D$. Let $F$ be the center of $D$. Suppose that $RF \subset R$, that is, $R$ is an algebra over $F$. Then $R$ satisfies a GPI over $F$ if and only if $R$ contains a minimal right ideal, and $D$ is finite-dimensional over $F$.

Proof. Let $C$ be the extended centroid of $R$. To prove the theorem, in view of Theorem 1.3.2, we merely must show that $C = F$.

Since $R$ is, by hypothesis, an algebra over $F$, $F \subset C$. So we must show that $C \subset F$; this is trivially equivalent to showing that $C \subset D$.

Let $c \neq 0 \in C$; hence there exists an ideal U of R such that $0 \neq c^{-1}U \subset R$. Because $0 \neq c^{-1}U$ is an ideal of R, and V is a faithful, irreducible R-module, $V = V(c^{-1}U)$. Hence, for some $v \in V$, $v(c^{-1}U) \neq 0$ and, being a submodule of V, $v(c^{-1}U) = V$.

Define $T_c: V \to V$ by $(v(c^{-1}u))T_c = vu$. To see that $T_c$ is well-defined, if $v(c^{-1}u) = 0$ we must have $vu = 0$. If $vu \neq 0$ then $vuU = V$ and so $vuU(c^{-1}U) \neq 0$ since $Uc^{-1}U \neq 0$ is an ideal of R. But $0 \neq vuUc^{-1}U = v(c^{-1}u)U^2 = 0$ since c centralizes R. This contradiction shows that $T_c$ is well-defined; it is clearly in D since c centralizes R, hence is in $F = Z(D)$.

Another immediate consequence of Martindale's theorem is an extremely important result due to Posner. But first we make the

DEFINITION. Let $A \subset B$ be rings. We say that A is a <u>right order</u> in B (or B is a <u>classical ring of right quotients</u> of A) if:

1. every regular element in A is invertible in B
2. if $x \in B$ then $x = ab^{-1}$ where $a, b \in A$ and b is regular.

We can similarly define a <u>left-order</u>

Now to Posner's theorem [25, 13].

THEOREM 1.3.4. Suppose that R is a prime ring and satisfies a polynomial identity over its centroid $\mathfrak{Z}$. Then R is a left and right order in $S = RC$, the central closure of R, and S is a finite-dimensional simple algebra over C with $C = Z(S)$.

Proof. As usual, we may assume that R satisfies a multilinear, homogeneous polynomial identity. Since R is an algebra over $\mathfrak{Z}$, we have $\mathfrak{Z} \subset C$, the extended centroid of R. Therefore $S = RC$ satisfies the same identity as does R. By Theorem 1.3.2, S must be a primitive ring; being a primitive ring satisfying a polynomial identity, by Kaplansky's theorem (see [14, 18, 20]), S is a simple algebra finite-dimensional over its center. Moreover, its center is precisely C. So, to prove Posner's theorem, all we need is to show that R is an order in S.

Let $U \neq 0$ be an ideal of R; then $UC \neq 0$ is an ideal of S. Since S is simple, we must have $S = UC$. However, since S is a finite-dimensional simple algebra, S must be isomorphic to $D_n$, the $n \times n$ matrices over a division algebra D. Thus the matrix units $e_{ii}$, being in S, must be in UC, whence $e_{ii} = \sum_j r_{ij}c_{ij}$ for suitable $r_{ij} \in U$, $c_{ij} \in C$.

Because the $c_{ij} \in C$, there is an ideal V of R such that $0 \neq c_{ij}V \subset R$ for all the i, j; we may suppose that $V \subset U$, since $V \cap U \neq 0$. Thus

$$e_{ii}V^3e_{ii} \subset (\sum r_{ij}c_{ij})V \cdot V \cdot V(\sum r_{ij}c_{ij})$$

$$\subset (\sum r_{ij}(c_{ij}V))V(\sum c_{ij}Vr_{ij}) \qquad (\text{since } c_{ij} \in C)$$

$$\subset RVR \subset V .$$

Since $e_{ii}V^3 \neq 0$ is a right ideal of R, it is not nilpotent, therefore $e_{ii}V^3e_{ii} \neq 0$. So U contains a non-zero element in $e_{ii}Se_{ii}$. But

$$e_{ii}Se_{ii} = \begin{pmatrix} 0 & & 0 \\ & D & \\ 0 & & 0 \end{pmatrix}, \qquad \text{so for some } d_i \neq 0 \in D,$$

$$\begin{pmatrix} 0 & & & & \\ & 0 & & & \\ & & d_i & & \\ & & & 0 & \\ & & & & 0 \end{pmatrix} \in V \subset U, \text{ for each } i. \text{ Thus } \quad u = \begin{pmatrix} d_1 & & & \\ & d_2 & & \\ & & \ddots & \\ & & & d_n \end{pmatrix} \in U.$$

Since $u$ is invertible in $S$ it is regular in $R$. In short, $U$ contains a regular element.

If $c_1, \ldots, c_k \in C$ we claim that there is an element $b \in R$, $b$ regular, such that $c_i = a_i b^{-1}$. For, there is an ideal $U \subset R$ with $0 \neq c_i U \subset R$ for each $i$. Since $U$ is an ideal of $R$, by the above, $U$ contains a regular element $b$, where $b^{-1}$ exists in $S$. So $c_i b = a_i \in R$ giving us $c_i = a_i b^{-1}$. In particular, if $x \in S$, since $x = \sum r_{ij} c_{ij}$, $r_{ij} \in U$, $c_{ij} \in C$ we have $c_{ij} = a_{ij} b^{-1}$ for $a_{ij} \in R$, $b$ regular in $R$. Thus $x = \sum r_{ij} a_{ij} b^{-1} = (\sum r_{ij} a_{ij}) b^{-1} = ub^{-1}$ where $u \in U$, $b$ regular in $R$.

Finally, if $t \in R$ is regular, then $t$ is regular in $S$. For, if $tx = 0$ for some $x \neq 0 \in S$, then since $x = ab^{-1}$, $a, b \in R$, we have $0 = tx = tab^{-1}$ giving $ta = 0$, and so, $a = 0$ since $t$ is regular in $R$. But a regular element in $S$, which is a finite-dimensional algebra, must be invertible in $S$. Hence $t$ regular in $R$ implies that $t$ is invertible in $S$. All the pieces of the theorem have been proved, and the theorem is now established.

We shall sharpen Posner's theorem considerably, in the next section. There it will follow from Formanek's theorem that if $R$ is a prime ring satisfying a polynomial identity over $\mathfrak{z}$, then the center of of R, $Z(R)$, is not 0 and every element in $S = RC$ is of the form $x = ab^{-1}$ where b is in $Z(R)$. So R will sit as an order in S in a particularly nice way.

## 4. Central Polynomials

An easy and well-known fact about the ring of $2 \times 2$ matrices over a field is that for any two such matrices x and y, $(xy - yx)^2$ must be a scalar. In fact this is true for any algebra which is quadratic over a field. Kaplansky asked if an analogous statement could be made for the ring of $n \times n$ matrices over a field. More precisely, Kaplansky's question was: if $F_n$ is the ring of $n \times n$ matrices over the field F, is there a polynomial $p(x_1, \ldots, x_k)$ which takes all its values in F, and does not take on a constant value ?

DEFINITION. A polynomial, with constant term 0, $p(x_1, \ldots, x_k) \in F[x_1, \ldots, x_k]$, in k non-commuting variables $x_1, \ldots, x_k$, is said to be a <u>central polynomial</u> on $F_n$ if $p(a_1, \ldots, a_k) \in F$ for all $a_1, \ldots, a_k \in F_n$, and $p(b_1, \ldots, b_k) \neq 0$ for some $b_1, \ldots, b_k \in F_n$.

In the terms just defined, Kaplansky's question becomes: do central polynomials exist on $F_n$ for every n? In a very beautiful and simple argument, Formanek [8] showed that the answer to this question is yes. Aside from its own independent interest, this result of Formanek is even more important because of what it implies. As we shall see, an easy consequence of Formanek's theorem will be that any semi-prime ring satisfying a polynomial identity must have a non-trivial center. Furthermore, if this ring R is prime — the situation prevailing in Posner's theorem — and if $Q_0$ denotes the ring of quotients of R, then $Z(Q_0) = C$ is the field of quotients of $Z(R)$, and $Q_0 = RC$. This tells us that any element in $Q_0$ is of the form $a\lambda^{-1}$ where $a \in R$, $\lambda \in Z(R)$. An even larger area of application of Formanek's theorem is in the interrelation of rings satisfying a polynomial identity and Azumaya algebras. Via Formanek's theorem we can bring a powerful result of M. Artin [6, 26] into play in the study of polynomial identity rings. Such applications can be found in Procesi's book [27], and in papers by Amitsur [3], Goldie [9], Herstein and Small [15], Rowen [30], and Small [32]. These give a representative idea of how this result of Formanek can be exploited. Razmyzlov [28] has given different central polynomials for $F_n$.

We now prove Formanek's theorem.

THEOREM 1.4.1. For every $n \geq 1$ there exists a central polynomial on $F_n$.

<u>Proof.</u> If $x_1, \ldots, x_{n+1}$ are n+1 commuting variables and $X, Y_1, \ldots, Y_n$ n+1 non-commuting variables, consider the F-linear map $\psi: F[x_1, \ldots, x_{n+1}] \to F[X, Y_1, \ldots, Y_n]$ defined by $\psi(x_1^{a_1} x_2^{a_2} \cdots x_{n+1}^{a_{n+1}}) = X^{a_1} Y_1 X^{a_2} Y_2 \cdots X^{a_n} Y_n X^{a_{n+1}}$. If $g(x_1, \ldots, x_{n+1}) \in F[x_1, \ldots, x_{n+1}]$, we evaluate $G(X, Y_1, \ldots, Y_n) = \psi(g(x_1, \ldots, x_{n+1}))$ on $F_n$. We do so for a particular choice of matrices in $F_n$, namely

$$X = \begin{pmatrix} \lambda_1 & & 0 \\ & \ddots & \\ 0 & & \lambda_n \end{pmatrix},$$

a diagonal matrix, and $Y_1 = e_{i_1j_1}, \ldots, Y_n = e_{i_nj_n}$ matrix units. To do so it is enough to know the outcome of evaluating any monomial $X^{a_1}Y_1X^{a_2}Y_2 \cdots Y_nX^{a_{n+1}}$ at the indicated substitutions for X and the $Y_i$. This evaluation is easily seen to be $\lambda_{i_1}^{a_1}\lambda_{i_2}^{a_2}\cdots\lambda_{i_n}^{a_n}\lambda_{j_n}^{a_{n+1}} e_{i_1j_1}e_{i_2j_2}\cdots e_{i_nj_n}$. Hence we get for these X and $Y_i$'s, $G(X, Y_1, \ldots, Y_n) = g(\lambda_{i1}, \ldots, \lambda_{in}, \lambda_{j_n})e_{i_1j_1}\cdots e_{i_nj_n}$. We now use a particular g, namely

$$g(x_1, \ldots, x_{n+1}) = \prod_{2<j<k<n}(x_j - x_k)^2 \prod_{2\le i\le n}(x_1 - x_i)(x_{n+1} - x_i).$$

Clearly $g(\lambda_{i_1}, \ldots, \lambda_{i_n}, \lambda_{j_n}) = 0$ unless $i_1, \ldots, i_n$ is a permutation of $1, 2, \ldots, n$ and $j_n = i_1$. But then $g(\lambda_{i_1}, \ldots, \lambda_{i_n}, \lambda_{j_n})$ becomes $D = \prod_{1\le i<j\le n}(\lambda_i - \lambda_j)^2$, the discriminant of $\lambda_1, \ldots, \lambda_n$. Here, then, $G(X, e_{i_1j_1}, \ldots, e_{i_nj_n}) = De_{i_1}e_{i_1}$ if $j_1 = i_2$, $j_2 = i_3$, etc. — a so-called ladder; otherwise $G(X, e_{i_1j_1}, \ldots, e_{i_nj_n}) = 0$. Let $H(X, Y_1, \ldots, Y_n) = G(X, Y_1, \ldots, Y_n) + G(X, Y_n, Y_1, \ldots, Y_{n-1}) + \ldots + G(X, Y_2, \ldots, Y_n, Y_1)$, the sum of G's as we permute the Y's cyclically.

If $X = \begin{pmatrix} \lambda_1 & & 0 \\ & \ddots & \\ 0 & & \lambda_n \end{pmatrix}$ and $Y_1 = e_{i_1j_1}, \ldots, Y_n = e_{i_nj_n}$ then, from what we obtained above, we have that $H(X, Y_1, \ldots, Y_n) = DI$ or 0, according as the $e_{i_1j_1}, \ldots, e_{i_nj_n}$ form a "ladder" or do not. At any rate, we do have $H(X, Y_1, \ldots, Y_n) \in F$ for any diagonal matrix X and $Y_1, \ldots, Y_n$ matrix units. Since G, and so H, is linear in $Y_1, \ldots, Y_n$ we thus get $H(X, Y_1, \ldots, Y_n) \in F$ for <u>any diagonal matrix X and any</u> matrices $Y_1, \ldots, Y_n \in F_n$. For the argument to be valid, we don't need that X is diagonal, merely that X is <u>diagonalizable</u>.

Let $K = F(t_1, \ldots, t_{n^2})$ be the rational function field in $t_1, \ldots, t_{n^2}$ over F. If X is the generic matrix in $t_1, \ldots, t_{n^2}$, then X is diagonalizable, so $H(X, Y_1, \ldots, Y_n) \in K$ for any $Y_1, \ldots, Y_n \in F_n$, and X generic. Specializing the entries of X to elements of F we get that $H(X, Y_1, \ldots, Y_n) \in F$ for any $X, Y_1, \ldots, Y_n \in F_n$; thus $H(X, Y_1, \ldots, Y_n)$ is a central polynomial on $F_n$, provided we could show that it does not take on a constant value.

So we now show that H is non-trivial. By its form, H is linear in $Y_1, \ldots, Y_n$, so if X is a <u>fixed</u> matrix in $F_n$ then, if $H(X, Y_1, \ldots, Y_n) = 0$ for all $Y_1, \ldots, Y_n \in F_n$, by the linearity in $Y_1, \ldots, Y_n$, $H(X, Y_1, \ldots, Y_n) = 0$ for all $Y_1, \ldots, Y_n \in L_n$, L any extension of F. Pick $X \in F_n$ so its companion matrix has distinct roots $\lambda_1, \ldots, \lambda_n$, lying in some extension L of F. Since X is diagonalizable, we saw that $H(X, e_{12}, e_{23}, \ldots, 3_{n1}) = \prod_{i<j} (\lambda_i - \lambda_j)^2 \neq 0$. So H, in L and whence, in F takes on a non-zero value. H is thus a central polynomial on $F_n$.

Now we know that there are central polynomials on $F_n$ we could wonder how these central polynomials on $F_n$ behave on $F_{n-1}$. The answer is provided us in

LEMMA 1.4.1. If $f(x_1, \ldots, x_k)$ is a central polynomial on $F_n$ then $f(a_1, \ldots, a_k) = 0$ all $a_1, \ldots, a_k \in F_{n-1}$; that is, f is a polynomial identity on $F_{n-1}$.

<u>Proof.</u> We embed $F_{n-1}$ into $F_n$ by sending a in $F_{n-1}$ into $\begin{pmatrix} a & 0 \\ 0 & 0 \end{pmatrix}$ in $F_n$. Thus, by assumption, if $a_1, \ldots, a_k \in F_{n-1}$, then

$f\left(\begin{pmatrix} a_1 & 0 \\ 0 & 0 \end{pmatrix}, \ldots, \begin{pmatrix} a_k & 0 \\ 0 & 0 \end{pmatrix}\right) \in F$. But

$$f\left(\begin{pmatrix} a_1 & 0 \\ 0 & 0 \end{pmatrix}, \ldots, \begin{pmatrix} a_k & 0 \\ 0 & 0 \end{pmatrix}\right) = \begin{pmatrix} f(a_1, \ldots, a_k) & 0 \\ 0 & 0 \end{pmatrix} \in F;$$

in consequence, $f(a_1, \ldots, a_k) = 0$.

We now prove one of the consequences — which we indicated earlier — of Formanek's theorem.

THEOREM 1.4.2. If $R$ is a semi-prime ring satisfying a polynomial identity then $R$ has a non-trivial center.

Proof. Since $R$ is semi-prime, it is a subdirect product of prime rings $R_\alpha$. By Posner's theorem (Th. 1.3.4) each $R_\alpha$ is an order in an algebra $Q_\alpha$, which is a simple algebra finite-dimensional over $Z_\alpha$; in fact $[Q_\alpha : Z_\alpha] \leq [\frac{n}{2}]^2$ where $n$ is the degree of the polynomial identity of $R$. Also, $Q_\alpha = R_\alpha Z_\alpha$.

Pick $Q_{\alpha_0}$ so that $[Q_{\alpha_0} : Z_{\alpha_0}]$ is maximal for the $R_\alpha$'s giving us the representation of $R$ as a subdirect product. If $F$ is the algebraic closure of $Z_{\alpha_0}$ then, by Wedderburn's theorem, $Q_F = Q_{\alpha_0} \otimes_{Z_{\alpha_0}} F \approx F_m$ for some $m$. Let $f(X, Y_1, \ldots, Y_m)$ be the Formanek polynomial described in the proof of Theorem 1.4.1; since $f$ is a non-trivial central polynomial on $F_n$, by its form (linearity in $Y_1, \ldots, Y_m$) it is a non-trivial central polynomial on $Q_F$, hence on $Q_{\alpha_0}$ and, finally, on $R_{\alpha_0}$, since $Q_{\alpha_0} = R_{\alpha_0} Z_{\alpha_0}$. If $[Q_\beta : Z_\beta] < m^2$ then $f$ vanishes on $Q_\beta$ by Lemma 1.4.1. Thus, if $r_1, \ldots, r_{m+1} \in R$ then $f(r_1, \ldots, r_{m+1})$ maps into the center of each $R_\alpha$ and, for some choice of $r_1, \ldots, r_{m+1}$, does not map into 0 in $R_{\alpha_0}$. Thus f takes on values in the center of R, and, more importantly, must take on some non-zero value. Hence the center of R cannot be 0.

COROLLARY. If R is a semi-prime ring satisfying a polynomial identity, and if $U \neq 0$ is an ideal of R then $U \cap Z(R) \neq 0$.

Proof. Since R is semi-prime, so must U be semi-prime. Because U satisfies a polynomial identity, by Theorem 1.4.2, $Z(U) \neq 0$. By the Lemma 1.1.5, $Z(U) \subset Z(R)$. Hence $U \cap Z(R) \neq 0$.

We are now able to sharpen Posner's theorem (Th. 1.3.4) considerably, and to identify precisely the center of the ring of quotients of a prime ring satisfying a polynomial identity. This is

THEOREM 1.4.3. Let R be a prime ring satisfying a polynomial identity. If $S = RC$ is the extended centroid of R then every element in S is of the form $az^{-1}$ where $a \in R$, $z \neq 0$ is in the center of R. In particular, $Z(S) = C$ is the field of quotients of $Z(R)$. S, as before, is a central simple algebra finite-dimensional over C.

Proof. Let $0 \neq s \in S$; thus there is an ideal U of R such that $0 \neq sU \subset R$. However, U is a prime ring itself satisfying a polynomial identity; thus by the corollary to Theorem 1.4.2, $U \cap Z(R) \neq 0$. Hence for some $z \neq 0 \in Z(R)$, $sz = a \in R$; since $z \neq 0$ is in $Z(R)$, it is regular in R, hence invertible in S. Thus $s = az^{-1}$ as claimed.

If, furthermore, $s \in Z(S) = C$, it is trivial to see that $a \in Z(R)$. Thus C is the field of quotients of $Z(R)$.

## 5. The Amitsur-Levitzki Theorem

Recall that the standard identity, $s_k(x_1, \ldots, x_k)$ of degree $k$ is defined by

$$s_k(x_1, \ldots, x_k) = \sum_{\sigma \in S_k} (-1)^\sigma x_{\sigma(1)} \cdots x_{\sigma(k)}$$

where $\sigma$ runs over $S_k$, the symmetric group of degree $k$ and where $(-1)^\sigma$ is $+1$ or $-1$ according as $\sigma$ is even or odd.

It is quite easy to show that if $F$ is a field (or, in fact, a commutative ring with 1) then $F_n$, the ring of $n \times n$ matrices over $F$, satisfies $s_{n^2+1}$ and satisfies no polynomial identity of degree less than $2n$ (see [14]). It is a very beautiful and important theorem due to Amitsur and Levitzki that $F_n$ does indeed satisfy $s_{2n}(x_1, \ldots, x_{2n})$. There have been many proofs given of this central result — see [4], [21], [27], [34] — but there is no doubt that a recent proof given by Rosset [29] is both the simplest and most natural. Whereas the other proofs given used some type of combinatorial argument, that of Rosset is purely algebraic and revealing. What it really does is to exhibit the Amitsur-Levitzki theorem as a direct consequence of the Cayley-Hamilton theorem. We now present Rosset's proof.

From the nature of the standard identity, it is clear that to prove that $F_n$ satisfies $s_{2n}(x_1, \ldots, x_{2n})$ it is sufficient to do so where $F$ is a field of characteristic 0. So we assume that char $F = 0$. Let $V$ be a vector space of dimension $2n$ over $F$ and let $E$ be the exterior algebra of $V$. Thus $R$, the even part of $E$, is a commutative ring and then $E_n \supset R_n$.

Let $v_1, \ldots, v_{2n}$ be a basis of V, where $v_i \wedge v_j = -\delta_{ij} v_j \wedge v_i$ in E. If $A_1, \ldots, A_{2n} \in F_n$ and $B = A_1 v_1 + \ldots + A_{2n} v_{2n} \in E_n$ then $B^{2n} = s_{2n}(A_1, \ldots, A_{2n}) v_1 \wedge v_2 \wedge \ldots \wedge v_{2n}$. The Amitsur-Levitzki theorem is thus equivalent to the statement that $B^{2n} = 0$.

To show that $B^{2n} = 0$, since $B^2 \in R_n$ and R is an algebra over the rational field, by the Cayley-Hamilton theorem and the Newton identities, it is enough to show that trace $B^{2i} = 0$ for all i.

But what is $B^{2i}$? From the form of B we see that $B^{2i} = \sum s_{2i}(A_{j_1}, \ldots, A_{j_{2i}}) v_{j_1} \wedge \ldots \wedge v_{j_{2i}}$. Since $v_{j_1} \wedge \ldots \wedge v_{j_{2i}} \in R$, to show that trace $B^{2i} = 0$ it is sufficient to prove that trace $s_{2i}(A_1, \ldots, A_{2i}) = 0$ for any matrices $A_1, \ldots, A_{2i} \in F_n$. Now, formally

$$s_{2i}(x_1, \ldots, x_{2i}) = x_1 s_{2i-1}(x_2, \ldots, x_{2i}) + x_2 s_{2i-1}(x_1, x_3, \ldots, x_{2i}) + \ldots .$$

On the other hand, since the permutation $(1, 2, \ldots, 2i)$ is <u>odd</u>,

$$s_{2i}(x_1, \ldots, x_{2i}) = -s_{2i-1}(x_2, \ldots, x_{2i})x_1 - s_{2i-1}(x_1, x_3, \ldots, x_{2i})x_2 - \ldots ,$$

whence

$$2s_{2i}(x_1, \ldots, x_{2i}) = x_1 s_{2i-1}(x_2, \ldots, x_{2i}) - s_{2i-1}(x_2, \ldots, x_{2i})x_1$$
$$+ x_2 s_{2i-1}(x_1, x_3, \ldots, x_{2i}) - s_{2i-1}(x_1, x_3, \ldots, x_{2i})x_2 + \ldots .$$

Therefore, $2s_{2i}(A_1, \ldots, A_{2i})$ is a sum of additive commutators, hence its trace is 0, and so trace $s_{2i}(A_1, \ldots, A_{2i}) = 0$. This completes the proof. Hence we have shown

THEOREM 1.5.1. If C is a commutative ring then $C_n$ satisfies $s_{2n}(x_1, \ldots, x_{2n})$, the standard identity of degree 2n.

## 6. Centralizers

In a recent series of papers, the structures of a ring R and that of the centralizer, $C_R(a)$, of the element $a \in R$ have been shown to be intimately related. The first paper in this series, by Herstein and Neumann [16], considers what happens to R if $C_R(a)$ is assumed to be simple, or semi-simple, artinian for certain types of elements a. Miriam Cohen [7] carried out a similar study when $C_R(a)$ is assumed to be a prime Goldie ring.

The result that we shall need later, while of this general flavor and philosophy, goes in a different direction. Here the concern is with what happens to R if $C_R(a)$ is assumed to satisfy a polynomial identity. The specific result we shall use is due to S. Montgomery [23]. Her theorem has been extended by M. Smith [33]; it is from Smith's paper that the general thread of our exposition will come.

We shall not develop this area of results to the point to which Montgomery and Smith have brought them. Instead, we shall concern ourselves only with a special case of these theorems which will be enough for our situation. They obtain results even when no separability condition is imposed, and for algebras over integral domains. For us it will be sufficient to work in algebras over fields, and with appropriate separability hypotheses.

We advise the interested reader who wishes to pursue these interesting results further to consult the papers by Montgomery and Smith cited above, and their joint paper [24]. There is also a closely related paper by Rowen [31].

Some closely related results were also obtained by Harchenko.

We make the

DEFINITION. Let $A$ be an algebra over a field $F$, if $a \in A$ is algebraic over $F$ then $a$ is said to be <u>separable</u> if its minimal polynomial over $F$ has no repeated roots.

We begin with an easy result.

LEMMA 1.6.1. Let $R$ be a simple algebra, finite dimensional over its center $Z$, with $\dim_Z R = r^2$. If $a \in R$ is separable over $Z$, and algebraic of degree $n$, and if $C_R(a)$ satisfies a polynomial identity of degree $m$, then $r \leq \frac{mn}{2}$. Also, $R$ satisfies the standard identity of degree $mn$.

Proof. We first show that we may assume that $R$ is the ring of $r \times r$ matrices over a field. Let $F$ be the algebraic closure of $Z$ and let $R_1 = R \otimes_Z F$; by Wedderburn's theorem, $R_1 \approx F_r$. In $R_1$, $C_{R_1}(a \otimes 1) = C_R(a) \otimes F$; since we may assume that $C_R(a)$ satisfies a multilinear identity of degree $m$, we have that $C_{R_1}(a)$ satisfies this identity of degree $m$. Hence the conditions on $a$ in $R$ carry over to $a \otimes 1$ in $R_1$. Thus, without loss of generality, $R = F_r$ where $F$ is algebraically closed.

Since $a$ is separable, its minimal polynomial is $m(x) = (x - \lambda_1) \cdots (x - \lambda_n)$ where the $\lambda_i \in F$ are distinct. We can bring $a$ to the diagonal form

$$a = \begin{pmatrix} \lambda_1 I_1 & & 0 \\ & \ddots & \\ 0 & & \lambda_n I_n \end{pmatrix}$$

where $I_i$ is the identity matrix of size $r_i$. From this it is clear that

$$C_{F_r}(a) = \begin{pmatrix} F_{r_1} & & O \\ & \ddots & \\ O & & F_{r_n} \end{pmatrix}$$

hence, by the Amitsur-Levitzki theorem (Theorem 1.5.1), $C_{F_r}(a)$ satisfies the standard identity of degree $2k$, where $k = \max r_i$, and no identity of lower degree. Therefore $2k \leq m$.

Since $r = \sum r_i \leq nk \leq \frac{nm}{2}$, we have that part of the lemma. By the Amitsur-Levitzki theorem, $C_{F_r}(a)$ satisfies the standard identity of degree $nm$.

Let $R$ be an algebra over a field $F$, and suppose that $a \in R$ is algebraic over $F$. If $\delta$ is the inner derivation defined by $a$ on $R$, that is, if $\delta(x) = ax - xa$ for all $x \in R$, then it is very easy to show that $\delta$ is algebraic over $F$.

Let $f(\lambda)$ be the minimal polynomial of $\delta$ over $F$. We wish to interrelate the separability of $a$ with properties of this polynomial $f(\lambda)$.

LEMMA 1.6.2. If $a$ is separable over $F$ then $f'(0) \neq 0$, where $f'$ is the derivative of $f$. Conversely, if $R$ is semi-prime and the minimal polynomial of $a$ factors into linear factors in $F$ and if $f'(0) \neq 0$ then $a$ is separable over $F$.

Proof. Suppose that $a$ is separable over $F$. By tensoring with the algebraic closure of $F$ we can reach a situation where the minimal polynomial of $a$ factors and the situation is not disturbed. So we may assume that the minimal polynomial of $a$ has its roots $\alpha_1, \ldots, \alpha_n$ distinct and in $F$. We can decompose $R$ as $R = \sum_{i,j=1}^{n} R_{ij}$, where $R_{ij} = \{x \in R \mid ax = \alpha_i x,\ xa = \alpha_j x\}$. To show that $f'(0) \neq 0$ we merely need show that $\delta^2(r) = 0$ forces $\delta(r) = 0$.

Suppose then that $\delta^2(r) = 0$. We can write $r = \sum r_{ij}$, $r_{ij} \in R$, whence $\delta(r) = \sum (\alpha_i - \alpha_j)r_{ij}$, $0 = \delta^2(r) = \sum (\alpha_i - \alpha_j)^2 r_{ij}$. Since the sum of the $R_{ij}$'s is direct, we get that $r_{ij} = 0$ if $i \neq j$. But then $\delta(r) = \sum (\alpha_i - \alpha_j)r_{ij} = 0$.

Now for the other half. Suppose that $R$ is semi-prime and that the minimal polynomial, $p(\lambda)$, of $a$ splits into linear factors in $F$. If $\alpha \in F$ is a multiple root of $p(\lambda)$, let $q(\lambda) = \dfrac{p(\lambda)}{(\lambda - \alpha)^2}$. Since $q(a)(a - \alpha) \neq 0$, and $R$ is semi-prime, there is an $r \in R$ such that $q(a)(a - \alpha)r\, q(a)(a - \alpha) \neq 0$. Let $s = q(a)rq(a)(a - \alpha)$; then $(a - \alpha)^2 s = 0$, $(a - \alpha)s \neq 0$ and $s(a - \alpha) = 0$. Hence $\delta(s) = as - sa = (a - \alpha)s - s(a - \alpha) = (a - \alpha)s \neq 0$ whereas $\delta^2(s) = \delta(\delta(s)) = \delta((a - \alpha)s) = (a - \alpha)^2 s = 0$. This says that $f'(0) = 0$. With this the lemma is proved.

To obtain the theorem we want we need to know the relationship between $C_R(a)$ and $C_{\overline{R}}(\overline{a})$ where $\overline{R}$ is a homomorphic image of $R$.

We assume throughout that $R$ is an algebra over the field $F$, that $a \in R$ is algebraic over $F$, that $\delta$ is the inner derivation induced by $a$ on $R$, and that $f(\lambda)$ is the minimal polynomial of $\delta$ over $F$. Clearly, since $\delta(a) = 0$, $\lambda \mid f(\lambda)$.

LEMMA 1.6.3. Suppose that $f'(0) \neq 0$. Then, for any homomorphic image $\overline{R}$ of $R$, $C_{\overline{R}}(\overline{a}) = \overline{C_R(a)}$.

Proof. We use the notation $\bar{r}$ for the image of the element $r \in R$ in $\bar{R}$. Let I be the kernel of the map of R onto $\bar{R}$, and suppose that $\bar{r} \in C_{\bar{R}}(\bar{a})$. Thus $\delta(r) \in I$.

Now $f(\lambda) = \lambda g(\lambda)$ and $g(\lambda) = \lambda h(\lambda) + \beta$ with $\beta \in F$. Since $\beta = g(0) = f'(0)$, by assumption $\beta \neq 0$. Hence $1 = -\beta^{-1}h(\lambda)\lambda + \beta^{-1}g(\lambda) = \lambda h_1(\lambda) + \mu g(\lambda)$, where $h_1 = -\beta^{-1}h(\lambda)$, $\mu = \beta^{-1}$. Thus $r = (\delta h_1(\delta))(r) + (\mu g(\delta))(r) = r_1 + r_2$ where $r_1 = \delta h_1(\delta)r$ and $r_2 = \mu g(\delta)(r)$. Because $\delta(r) \in I$, we have that $r_1 \in I$, whence $\bar{r} = \bar{r}_2$. But $\delta(r_2) = \mu\delta g(\delta)(r) = \mu f(\delta)(r) = 0$, which is to say, $r_2 \in C_R(a)$. Hence from $\bar{r} = \bar{r}_2$ we obtain $C_{\bar{R}}(\bar{a}) \subset \overline{C_R(a)}$. However, the other containment relation $C_{\bar{R}}(\bar{a}) \supset \overline{C_R(a)}$ is trivial. With this we have the lemma.

We have all the pieces to carry out the proof of the main result of this section.

THEOREM 1.6.1. Let R be an algebra over a field F and suppose that $a \in R$ is algebraic over F. Suppose that $\delta$ defined on R by $\delta(x) = ax - xa$ satisfies $f(\lambda)$ over F where $f'(0) \neq 0$. Then, if $C_R(a)$ satisfies a polynomial identity of degree m over F

1. if R is semi-prime, R must satisfy $S_{nm}$, the standard identity of degree nm,

2. otherwise, R satisfies $S_{nm}^k$ for some k.

Proof. We first dispose of the semi-prime case. If R is semi-prime, it is a subdirect product of prime rings $R_\alpha$. By the preceding lemma, the hypotheses on R transfer over to $R_\alpha$. So, to settle the semi-prime case it is enough to settle the prime case. Hence we assume that R is prime.

Let C be the extended centroid of R; C is a field and $F \subset C$. We look at RC, the central closure of R.

Using the notation used in the proof of the preceding lemma, $1 = \delta h_1(\delta) + \mu g(\delta)$ holds over C, for $\delta$ acting on RC. If $s = \sum r_i c_i \in RC$, with $r_i \in R$ and $c_i \in C$ then $s = \delta h_1(\delta)(s) + \mu g(\delta)s = 0 + \mu g(\delta)(s)$ if $s \in C_{RC}(a)$ Hence $s = \mu g(\delta)(s) = \sum (\mu g(\delta)(r_i))c_i \in C_R(a)C$ since $g(\delta)(r_i) \in C_R(a)$ for an $r_i \in R$. Thus $C_{RC}(a) = C_R(a)C$.

Let L be the algebraic closure of C and let $R_1 = RC \otimes_C F$. $R_1$ is its own central closure. If $f_1(\lambda)$ is the minimal polynomial for $\delta \otimes 1$ over L then $\lambda | f_1(\lambda)$ and $f_1(\lambda) | f(\lambda)$. From $f'(0) \neq 0$ we then get $f_1'(0) \neq 0$. By Lemma 1.6.2, $a \otimes 1$ must be separable over F. So we may assume that R is its own central closure and its extended centroid F is algebraically closed; moreover, we may suppose that a is separable over F.

Let $p(\lambda) = \prod_1^k (\lambda - \alpha_i)$ be the minimal polynomial of a over F, with the $\alpha_i$ distinct, and let $p_j(\lambda) = \prod_{i \neq j} (\lambda - \alpha_i)$. We can write $1 = \sum q_j(\lambda) p_j(\lambda)$ since the $p_j$'s are relatively prime. Let $e_j = q_j(a)p_j(a)$; the set $e_1, \ldots, e_k$ are mutually orthogonal idempotents, with sum 1, such that $e_j a = a e_j = \alpha_j e_j$. Hence $e_j R e_j \subset C_R(a)$, so must satisfy a polynomial identity of degree m. Moreover, the center of $e_j R e_j$ is $Fe_j$. For suppose that $x \in Z(e_j R e_j)$; then for $r \in R$, $x e_j r e_j = e_j r e_j x$. Because $x \in e_j R e_j$, $x e_j = e_j x = x$, whence $x r e_j = e_j r x$. By Lemma 1.3.2, $x = c e_j$ where $c \in F$, the extended centroid of R. Now, $e_j R e_j$ is prime and satisfies a polynomial identity and its center is $Fe_j$ — in fact the argument above can be used to show that $Fe_j$ is the extended centroid of R. By Theorem 1.4.3, $e_jRe_j$ is central simple over over $Fe_j$ and is of dimension at most $[\frac{m}{2}]^2$, since F is algebraically closed.

Because $e_jRe_j$ satisfies a polynomial identity, R satisfies a GPI, so by Theorem 1.3.2 (since R is its own central closure) R is primitive with minimal right ideal. If M is a faithful irreducible R-module then $Me_j$ is a faithful irreducible right $e_jRe_j$-module of dimension at most $[\frac{m}{2}]$ over F. Since $M = \sum Me_j$, $\dim_F M \leq k[\frac{m}{2}] \leq \frac{nm}{2}$. Hence R is central simple over F of dimension at most $\frac{nm}{2}$; by the Amitsur-Levitzki theorem, R satisfies $S_{nm}$. This finishes the semi-prime case.

To go to the general case we use an old trick of Amitsur. Let N be the lower nil radical of R — that is, N is the intersection of all the prime ideals of R. Then R/N is semi-prime, and by virtue of Lemma 1.6.2, inherits the hypothesis of R. So R/N satisfies $S_{mn}$. So, if $x_1, \ldots, x_{mn} \in R$ then $S_{mn}[x_1, \ldots, x_{mn}] \in N$. Let $d = mn$ and let $\overline{R}$ be the complete direct product of copies of R indexed by the set of all d-tuples of elements of R. So $\overline{R} = \prod_{\alpha \in I} R_\alpha$, $R_\alpha = R$, $I = \{(r_1, \ldots, r_d) \mid r_i \in R\}$. Choose $f_1, \ldots, f_d \in \overline{R}$ so that $f_i(\alpha) = r_i$ and let $f_a \in \overline{R}$ be given by $f_a(\alpha) = a$ for all $\alpha \in I$. Then $f_a$ is algebraic and separable over F, $C_{\overline{R}}(f_a) = \prod_{\alpha \in I} C_R(a)_\alpha$, so $C_{\overline{R}}(f_a)$ satisfies the polynomial identity of $C_R(a)$. Thus, by the above, $S_d[f_1, \ldots, f_d] \in N(\overline{R})$, the lower radical of $\overline{R}$. Therefore, for some integer k, $S_d[f_1, \ldots, f_d]^k = 0$. But this implies that $S_d^k$ is an identity for R. This completes the proof of the theorem.

A special case, but in fact the one we shall need, is Montgomery's result.

THEOREM 1.6.2. Let $R$ be any ring, $Z$ its center and suppose that $a \in R$ is such that $a^n \in Z$, where both $a$ and $n$ are invertible in $R$. If $C_R(a)$ satisfies a polynomial identity of degree $m$ then

1. if $R$ is semi-prime, $R$ satisfies $S_{mn}$,
2. otherwise, $R$ satisfies $S_{mn}^k$ for some $k$.

Proof. To prove the result, as we did in the proof of Theorem 1.6.1, it will suffice to do so for the prime case.

So, suppose that $R$ is prime; thus $Z$ is an integral domain. Let $F$ be the algebraic closure of the field of fractions of $Z$ and let $V = R \otimes_Z F$. $V$ is an algebra over the field $F$. Since $a$ is invertible in $R$, the map $\varphi$ defined on $V$ by $\varphi(v) = a^{-1}va$ makes sense (and is an automorphism of $V$). Since $a^n \in Z$, $\varphi^n = 1$. Since $n$ is invertible in $F$, the n-th roots of 1 are distinct and lie in $F$ (since $F$ is algebraically closed). So we can decompose $V$ as $V = C_V(a) \oplus \sum_{\lambda \neq 1} V_\lambda$ where $V_\lambda = \{v \in V \mid \varphi(v) = \lambda v\}$. If $\delta$ is defined on $V$ by $\delta(x) = ax - xa$ then for $v \in V_\lambda$, $\delta(v) = av - va = (1-\lambda)av$. Since $a$ commutes with $\delta$, this gives $\delta^n(v) = (1-\lambda)^n a^n v = (1-\lambda)^n zv$ where $z = a^n \in Z$. Thus $\delta \prod_{\lambda \neq 1} (\delta^n - (1-\lambda)^n z) = 0$ on $V$, hence $\delta$ is a root of $f(x) = x \prod_{\lambda \neq 1} (x^n - (1-\lambda)^n z)$ over $F$. But then $f'(0) = (\pm \prod_{\lambda \neq 1} (1-\lambda)^n) z^{n-1} \neq 0$. Since, as is easy to verify (assuming that the polynomial identity of $C_R(a)$ is multilinear) $C_V(a)$ satisfies the identity of $C_R(a)$, by applying Theorem 1.6.1, we have that $V$, and so $R$, satisfies $S_{mn}$. This proves the theorem.

Although we do not need the situation in which $f'(0)$ is not invertible, we state what is known here. These results come from the works of Montgomery and Smith. For $R$ prime, the story is a fairly decisive one, namely:

If $R$ is prime and $a \in R$ is algebraic over $Z$ of degree $n$, and if $C_R(a)$ satisfies $S_{2d}^k$ then $R$ satisfies $S_{2nkd}$.

To see that such results are false in general we present an example of George Bergman. Let $A$ be an algebra with $1$ over a field $F$ of characteristic $p \neq 0$; suppose that $A$ satisfies no polynomial identity. Let $R = \{\begin{pmatrix} a & b \\ 0 & \alpha \end{pmatrix} \mid a, b \in A, \alpha \in F\}$; clearly, since $R \supset \begin{pmatrix} A & 0 \\ 0 & 0 \end{pmatrix}$, $R$ does not satisfy a polynomial identity. Let

$a_0 = \begin{pmatrix} 1 & 1 \\ 0 & 1 \end{pmatrix} \in R$; then $a^p = 1$ and $C_R(a) = \{\begin{pmatrix} \alpha & b \\ 0 & \alpha \end{pmatrix} \mid \alpha \in F, b \in A\}$

$C_R(a)$ satisfies $(xy - yx)^2 z - z(xy - yx)^2$, yet $R$ satisfies no polynomial identity.

Bibliography

1. S. A. Amitsur. Generalized polynomial identities and pivotal polynomials. Trans. AMS 114 (1965): 210-226.

2. S. A. Amitsur. On rings of quotients. In Symposia Matematica VIII (Academic Press 1972).

3. S. A. Amitsur. Polynomial identities and Azumaya algebras. Jour. Algebra 26 (1973): 117-125.

4. S. A. Amitsur and J. Levitzki. Minimal identities for algebras. Proc. AMS 1 (1950): 449-463.

5. V. Andrunakievitch and J. M. Rjahubin. Rings without nilpotent elements, and completely prime ideals. Doklady Akad. Nauk USSR 180 (1968): 9-11.

6. M. Artin. On Azumaya algebras and finite dimensional representation of rings. Jour. Algebra 11 (1969): 532-563.

7. Miriam Cohen. Semi-prime Goldie centralizers. Israel Jour. Math. 20 (1975): 37-45.

8. E. Formanek. Central polynomials for matrix rings. Jour. Algebra 23 (1972): 129-133.

9. A. W. Goldie. On Artin's theorem. (to appear).

10. I. N. Herstein. Lie and Jordan structures in simple associative rings. Bull. AMS 67 (1961): 517-531.

11. I. N. Herstein. On the Lie structure of an associative ring. Jour. Algebra 14 (1970): 561-571.

12. I. N. Herstein. Sui commutatori degli anelli semplici. Rendiconti Sem. Milano XXXIII (1963): 3-9.

13. I. N. Herstein. Topics in Ring Theory. Univ. of Chicago Press, 1969.

14. I. N. Herstein. Non-Commutative Rings . Carus Monograph 15, 1968.

15. I. N. Herstein and Lance Small. The intersection of the powers of the radical in P. I. Noetherian rings. Israel Jour. Math. 16 (1973): 176-180.

16. I. N. Herstein and Linda Neumann. Centralizers in rings. Annali di Mat. CII (1975): 37-44.

17. N. Jacobson. On the theory of primitive rings. Annals of Math. 48 (1947): 8-21.

18. N. Jacobson. Structure of Rings. AMS Colloq. Publications XXXVII (2d ed.) 1964.

19. I. Kaplansky. Fields and Rings. Univ. of Chicago Press (2d ed.), 1972.

20. I. Kaplansky. Rings with polynomial identity. Bull. AMS 54 (1948): 575-580.

21. B. Kostant. A theorem of Frobenius , a theorem of Amitsur-Levitzki and cohomology theory. Jour. Math. Mech. 7 (1958): 237-264.

22. W. Martindale. Prime rings satisfying a generalized polynomial identity. Jour. Algebra 12 (1969): 576-584.

23. Susan Montgomery. Centralizers satisfying polynomial identities. Israel Jour. Math. 18 (1974): 207-219.

24. Susan Montgomery and Martha Smith. Algebras with a separable subalgebra whose centralizer satisfies a polynomial identity. Communications Algebra 3 (1975): 151-168.

25. E. Posner. Prime rings satisfying a polynomial identity. Proc. AMS 11 (1960): 180-184.

26. C. Procesi. On a theorem of M. Artin. Jour. Algebra 22 (1972): 309-315.

27. C. Procesi. Rings With Polynomial Identities, Dekker 1973.

28. J.P. Razmyslov. On a problem of Kaplansky. Mathematics of USSR, Izvestia 7 (1973): 479-496.

29. S. Rosset. On the Amitsur-Levitzki theorem. (to appear).

30. L. Rowen. The Artin-Procesi theorem revisited. (unpublished).

31. L. Rowen. Generalized polynomial identities II. (to appear).

32. Lance Small. Prime ideals in Noetherian P.I. rings. Bull AMS 79 (1973): 421-422.

33. Martha Smith. Rings with an integral element whose centralizer satisfies a polynomial identity. Duke Math. Jour. 42 (1975): 137-149.

34. R. Swan. An application of graph theory to algebra. Proc. AMS 14 (1963): 367-373.

35. Y. Utumi. On quotient rings. Osaka Math. Jour. 8 (1956): 1-18.

# 2

# REGULARITY CONDITIONS ON SKEW AND SYMMETRIC ELEMENTS

In the first chapter, although rings with involution did occasionally arise, the discussion was usually about general rings. Now we shall narrow our attention to rings with involution, and to particular types of problems about them. In most of what follows R will be a ring with involution *. For such rings we shall consistently use the notation $S = \{x \in R \mid x = x^*\}$ for the set of symmetric elements of R and $K = \{x \in R \mid x^* = -x\}$ for the set of skew elements of R. In addition, we shall often be concerned with the set of traces T in R defined by $T = \{x + x^* \mid x \in R\}$ and the set of skew-traces $K_0 = \{x - x^* \mid x \in R\}$ of R. Clearly $T \subset S$ and $K_0 \subset K$. Our concern with these special subsets of S and K stems from the fact that relative to a *-homomorphism of R the traces and skew-traces in the images ring are images of traces and skew-traces in the ring R itself. This need not be true for the symmetric and skew elements of the image ring; new symmetric and skew elements may arise in the image ring if it should pick up some 2-torsion.

We shall have many occasions to make use of results of ours about the Lie and Jordan structure of associative rings. Most of these results — at least, some version of them — can be found in our monograph Topics in Ring Theory [3], which will be referred to by the designation TRT. We shall re-do, or extend, some of these results here; we do this when we feel that we have either something new to add to the result as stated in TRT or a different way of looking at the result.

In order to have these results handily available we shall cite some of them, with a numbering consistent with the present text. In the immediate few pages that follow $R$ will be any ring, not necessarily one with involution.

An additive subgroup $A \subset R$ is called a Jordan subring of $R$ if $a, b \in A$ implies that $a \circ b = ab + ba \in A$. When $A$ is a Jordan subring of $A$, and $B \subset A$ is an additive subgroup such that $a \in A$, $b \in B$ implies $ab + ba \in B$, then $B$ is called a Jordan ideal of $A$. We might write this as: $B \subset A$ is a Jordan ideal of $A$ if $B \circ A \subset A$.

An additive subgroup $L \subset R$ is called a Lie subring of $R$ if whenever $a, b \in L$ then $[a, b] = ab - ba \in L$. If $L$ is a Lie subring of $R$, and if $U \subset L$ is an additive subgroup such that $au - ua \in U$ for any $a \in L$, $u \in U$, then we call $U$ a Lie ideal of $L$. If, for subsets $X, Y$ of $R$ we denote by $[X, Y]$ the additive subgroup generated by all $xy - yx$ where $x \in X$ and $y \in Y$, then we could restate that $U$ is a Lie ideal of $L$ by merely noting that this is equivalent to saying that $[U, L] \subset L$.

## 1. Osborn's Theorem.

We now begin the list of results from TRT which will be useful to us here.

LEMMA 2.1.1. If R is any ring and $\rho \neq 0$ is a nil right ideal then, if $\rho$ satisfies a polynomial identity, R has a non-zero nilpotent ideal. In particular, if every $x \in \rho$ satisfies $x^n = 0$ where n is a fixed integer, then R cannot be semi-prime.

THEOREM 2.1.1. Let R be a semi-prime, 2-torsion free ring and suppose that $U \neq 0$ is a Jordan ideal of R. Then U contains a non-zero ideal of R.

COROLLARY. If R is a simple ring of characteristic not 2, then R is simple as a Jordan ring.

This theorem and its corollary are the content of TRT Theorem 1.1.

The next result, as stated here, is not exactly that stated in TRT Lemma 1.3; however, a glance at that lemma and its proof reveals that what is proved is exactly what we state here.

THEOREM 2.1.2. Let R be any ring and suppose that A is both a subring of R and a Lie ideal of R. Then A contains the ideal $R[A, A]R$ of R. In particular, if R is semi-prime, then:

1. if A is not commutative, A contains a non-zero ideal of R.
2. if A is commutative, then if $a \in A$ we must have $a^2 \in Z(R)$.
3. if A is commuative and R is 2-torsion free, then $A \subset Z(R)$.

This result leads to (TRT Theorem 1.5)

THEOREM 2.1.3. If R is a simple ring and if U is a Lie ideal of R then either $U \subset Z(R)$ or $U \supset [R, R]$, except if R is of characteristic 2 and $\dim_Z R = 4$.

A sharper version of Theorem 2.1.3 is provided in (TRT Theorem 1.14)

THEOREM 2.1.4. If $R$ is a simple ring, and if $T$ is an additive subgroup of $R$ such that $[T,[R,R]] \subset T$ then either $T \subset Z(R)$ or $T \supset [R,R]$, except if $R$ is of characteristic 2 and $\dim_Z R = 4$.

A special case of Theorem 2.1.4, but one of great interest, is its

COROLLARY. If $R$ is a simple ring and if $U$ is a Lie ideal of $[R,R]$ then either $U \subset Z(R)$ or $U = [R,R]$, except if $R$ is of characteristic 2 and $\dim_Z R = 4$. That is, except for the exception noted, $[R,R]/[R,R] \cap Z(R)$ is a simple Lie ring.

This corollary is TRT Theorem 1.13.

We pause to give some classes of examples of Lie ideals in rings. Firstly, we have some obvious, trivial examples, namely, any ideal of the ring, the center of the ring, and $[R,R]$ and any additive subgroup which contains $[R,R]$. However, there are less obvious and more interesting examples.

<u>Example 1.</u> Let $E$ be the additive subgroup of $R$ generated by the idempotents of $R$. If $e^2 = e$ is an idempotent in $R$, and if $x \in R$, then it is a simple calculation to verify that $f = e + ex - exe$ and $g = e + xe - exe$ are idempotents. Hence $ex - xe = f - g$ is in $E$. This gives us that $E$ is a Lie ideal of $R$.

In particular, using the theorems cited above, one can show that in any simple ring having an idempotent $e \neq 0, 1$, $E$ must contain $[R,R]$. One might wonder how close $E$ comes to being all of $R$. For this to make sense, even for matrices, we must look at a set larger than E. In matrices the trace of an idempotent is an integer, hence the trace of any element in E is also an integer. Thus one cannot expect that E = R. Instead of using E, let $\mathcal{E}$ be the vector space generated by all the idempotents of R over $\mathfrak{Z}$, the centroid of R. For matrices, it then is true that $\mathcal{E} = R$. The following is open:

Question: if R is simple and R contains an idempotent $e \neq 0, 1$, is $\mathcal{E} = R$?

Example 2. Let R be any ring with involution * and let $\overline{S}$ be the subring of R generated by S, the set of symmetric elements of R. If $s \in S$, $r \in R$ then $sr - rs = (sr + r^*s) - (r + r^*)s$, so is in $\overline{S}$. Since commutation is a derivation of R, we immediately get that $[\overline{S}, R] \subset \overline{S}$, hence $\overline{S}$ is a Lie ideal of R.

As is clear from the proof just given, $\overline{T}$, the subring generated by T, the set of traces in R, is also a Lie ideal of R.

We see some consequences of this general fact that $\overline{S}$ is always a Lie ideal of R.

THEOREM 2.1.5. Let R be a semi-prime ring with involution. Then either $S \subset Z$, the center of R, or $\overline{S}$ contains a non-zero ideal of R. In the former case, R must satisfy the standard identity of degree 4.

Proof. Since $\overline{S}$ is both a subring and a Lie ideal of R, if $\overline{S}$ is not commutative we have by Theorem 2.1.2 that $\overline{S}$ contains a non-zero ideal of R. So we may suppose that $\overline{S}$ is commutative, in which case $\overline{S}$ is merely S. Also from Theorem 2.1.2 we know that if R is

2-torsion free then $S \subset Z$, and, in any event, if $a \in S$ then $a^2 \in Z$.

Let $W = \{x \in R \mid 2x = 0\}$; $W$ is an ideal of $R$, is of characteristic 2, and $W^* \subset W$. Let $S_W = S \cap W$; if $a \in S_W$ and $x \in W$ then $a(x+x^*) = (x+x^*)a$, hence $ax + xa = ax^* + x^*a = (ax + xa)^*$. That is, $ax+xa \in S_W$, and commutes with $a$.

Now, $axx^* = xx^*a$; thus $(ax+xa)x^* = x(ax^* + x^*a) = x(ax+xa)$, since $ax+xa = ax^* + x^*a$. This tells us that $x(ax+xa) \in S_W$, hence must commute with $a$; from $ax(ax+xa) = x(ax+xa)a = xa(ax+xa)$ we get that $(ax+xa)^2 = 0$. Thus if $a \in S_W$, $x \in W$ then $(ax+xa)^2 = 0$.

Suppose that $a \in S_W$ and $a \notin Z(S_W)$; then $b = ax + xa \neq 0$ for some $x \in W$. By the above, $b^2 = 0$ and $(by + yb)^2 = 0$ for all $y \in W$. Thus $0 = b(by + yb)^2 y = (by)^3$. Since $R$ is semi-prime and $W$ is an ideal of $R$, $W$ is semi-prime. By Lemma 2.1.1, $(by)^3 = 0$ for all $y \in W$ forces $b = 0$. With this we have that $S_W \subset Z(S_W) \subset Z$, by Lemma 1.1.5.

Let $\tilde{R} = R/W$; since $R$ is semi-prime, $\tilde{R}$ is 2-torsion free, and semi-prime. Let $\tilde{x}^* = \tilde{x}$, $\tilde{y}^* = \tilde{y}$ in $\tilde{R}$ and let $x, y$ be inverse images in $R$ of $\tilde{x}, \tilde{y}$ respectively. Thus $x^* - x \in W$, $y^* - y \in W$ hence $2x^* = 2x$, $2y^* = 2y$; since $S$ is commutative, we have $(2x)(2y) = (2y)(2x)$, that is, $4(xy - yx) = 0$. Because $R$ is semi-prime, we have $2(xy - yx) = 0$, which translates in $\tilde{R}$ into $\tilde{x}\tilde{y} = \tilde{y}\tilde{x}$. Because $\tilde{R}$ is semi-prime, 2-torsion free, and any two of its symmetric elements commute, all symmetric elements of $\tilde{R}$ are in $Z(\tilde{R})$.

If $s \in S$ then $\tilde{s}^* = \tilde{s}$, hence we have $sx - xs \in W$ for all $x \in R$. But $s(x+x^*) = (x+x^*)s$ gives us that $sx - xs \in S$ thus $sx - xs \in S_W \subset Z(W) \subset Z$. Also, $Z \ni s(sx) - (sx)s = s(sx - xs)$, therefore $(s(sx - xs))x = x(s(sx - xs))$. Because $sx - xs \in Z$ this gives us $(sx - xs)^2 = 0$. However, $sx - xs$ is in $Z$ and is nilpotent, which can happen in the semi-prime ring $R$ only if $sx - xs = 0$. Thus $S \subset Z$ has been proved.

If $x \in R$ then $x^2 - (x+x^*)x + x^*x = 0$ and since $x + x^* = \alpha \in Z$, $x^*x = \beta \in Z$, $x$ is quadratic over $Z$. Thus $R$ satisfies the polynomial identity $[[x^2, y], [x, y]] = 0$ of degree 5. Because $R$ is semi-prime we then easily get that $R$ satisfies the standard identity of degree 4.

The theorem has the following consequence (Theorem 1.6 in TRT)

THEOREM 2.1.6. Let $R$ be a simple ring with involution. If $\dim_Z R > 4$ then $\overline{S} = R$, that is, $S$ generates $R$.

Proof. By the preceding theorem, $\overline{S}$ contains a non-zero ideal of $R$, in which case $\overline{S} = R$, or $R$ satisfies the standard identity of degree 4. This second possibility, however, implies that $\dim_Z R \leq 4$. This proves Theorem 2.1.6.

We now go on to prove a generalization of a very interesting theorem due to Osborn [12]. Osborn proved the result for simple rings, od characteristic not 2, assuming every non-zero symmetric element to be invertible. We shall prove the result for any semi-prime ring in which we insist that only the non-zero traces be invertible. We first, however, handle the case in which all the symmetric elements are invertible.

THEOREM 2.1.7. Let R be a semi-prime ring in which every symmetric element $s \neq 0$ is invertible in R. Then R is

1. a division ring, or

2. the direct sum of a division ring and its opposite, relative to the exhange involution $(x, y)^* = (y, x)$, or

3. the $2 \times 2$ matrices over a field, relative to the symplectic involution, namely, $\begin{pmatrix} \alpha & \beta \\ \gamma & \delta \end{pmatrix}^* = \begin{pmatrix} \delta & -\beta \\ -\gamma & \alpha \end{pmatrix}$.

<u>Proof.</u> If $A \neq 0, R$ is an ideal of R such that $A^* = A$ then A contains no invertible elements, hence A can have no non-zero symmetric elements. However, if $0 \neq x \in A$ then $x^* \in A$ and $t = x + x^* \in A$ is symmetric. In consequence, $t = 0$, and $x^* = -x \neq 0$. Thus A cannot have any 2-torsion, otherwise A would contain non-zero symmetric elements. If $x \in A$ then $xx^* \in A$, and is symmetric; thus $xx^* = 0$. But, since $x^* = -x$, this gives us that $0 = xx^* = -x^2$, so that $x^2 = 0$ for all $x \in A$. Since A is 2-torsion free we get, from the fact that all its elements have square 0, that $A^3 = 0$, contradicting the semi-primeness of R. Hence we may assume that R has no proper non-zero *-ideals.

Suppose that $I \neq 0$, R is an ideal of R. Then, since $I + I^*$ is a non-zero *-ideal of R, $I + I^* = R$. Also, $I \cap I^* \neq R$ is a *-ideal of R, hence $I \cap I^* = 0$. In short, $R = I \oplus I^*$. We claim that I is a division ring. For, $1 = e + f^*$, $e, f \in I$ gives us that e is the unit element of I. Also, if $x \neq 0$ is in I, then $x + x^* \neq 0$ is invertible in R; if its inverse is $y_1 + y_2^*$ where $y_1, y_2 \in I$, we get that $xy_1 = e$. Thus I is a division ring, as is $I^*$. Clearly, the involution on R is the exchange involution.

Suppose, finally, that $R$ is simple but not a division ring. Let $0 \neq x \in R$ be non-invertible in $R$. Since $xSx^* \subset S$ and no element of $xSx^*$ is invertible, $xSx^* = 0$. Thus, if $r \in R$ then $x(r+r^*)x^* = 0$, that is, $xrx^* = -xr^*x^* = -(xrx^*)^*$. Since $R$ is simple, $k = xrx^* \neq 0$; then $k^* = -k \neq 0$ is not invertible $(k^2 = 0)$ in $R$. In particular, char $R \neq 2$.

Now $kSk \subset S$ and consists of non-invertible elements. Hence $kSk = 0$. If $s \in S$ then $sk - ks \in S$; but

$(ks - sk)^2 = ksks + sksk - ks^2k - sk^2s = 0$ since $kSk = 0$ and $k^2 = 0$.

Thus $ks - sk = 0$. Thus $k$ centralizes $S$, hence $\overline{S}$. Since $k \notin Z$ (being nilpotent), $\overline{S} \neq R$. By Theorem 2.1.6 we get that $\dim_Z R = 4$; since $R$ is <u>not</u> a division ring, $R \approx F_2$ for a field $F$.

From the nature of theinvolutions on $F_2$ it is easy to see that the involution $*$ on $F_2$ must be the symplectic involution. This finishes the proof.

Note that in carrying out the proof we actually showed that possibility 3 cannot occur if char $R = 2$. So we have the

COROLLARY. If $R$ is a semi-prime ring of characteristic 2 with involution in which every non-zero symmetric element is invertible, then either $R$ is a division ring or the direct sum of a division ring and its opposite relative to the exchange involution.

We sharpen Theorem 2.1.7 by merely restricting the traces to be invertible [5].

THEOREM 2.1.8. Let $R$ be a semi-prime ring with involution in which every non-zero trace $x+x^*$ is invertible in $R$. Then $R$ is

1. a commutative ring of characteristic 2, with no non-zero nilpotent elements, in which $x^* = x$ for all $x \in R$, or

2. a division ring, or

3. the direct sum of a division ring and its opposite, relative to the exchange involution, or

4. the ring of $2 \times 2$ matrices over a field relative to the symplectic involution.

Proof. If $x+x^* = 0$ for all $x \in R$ then, since $x^* = -x$, $-x^2 = (x^2)^* = (x^*)^2 = x^2$, and so $2x^2 = 0$. Because R is semi-prime, we get that $2R = 0$, hence R is of characteristic 2. Therefore $x^* = x$ for all $x \in R$ and R is commutative. Being semi-prime and commutative, R has no nilpotent elements.

So we may suppose that $t = x+x^* \neq 0$ for some $x \in R$. If $A^* = A$ is an ideal of R, $A \neq R$ then $a+a^* = 0$ for all $a \in A$, for otherwise A would contain an invertible element. If $r \in R$ then $ra \in A$ for $a \in A$, hence $(ra)^* = -ra$; but $(ra)^* = a^*r^* = -ar^*$. Consequently, $ra = ar^*$ for all $r \in R$, $a \in A$. If $w \in R$, then $(wr)a = a(wr)^* = ar^*w^* = -raw^* = rwa$, in consequence of which $(wr - rw)A = 0$ follows. Since R is semi-prime, if $w \in A$ then $wr - rw \in A$ and in the annihilator of A, $wr - rw = 0$ must hold. Hence $A \subset Z$.

Now if $a \neq 0 \in A$, since $a^* = -a$, $(a^2)^* = a^2$; if $t = x+x^* \neq 0$ then $a^2t = a^2x + a^2x^* = a^2x + x^*a^2$ since $A \subset Z$, so is invertible or 0; being in A it is not invertible. Hence $a^2t = 0$ and so $a^2 = 0$ because t is invertible. In short, $a \in A$ implies $a^2 = 0$. In a semi-prime ring this is not possible unless $A = 0$. Hence the only *-ideals of R are 0 and R.

As in the proof of the preceding theorem, if $R$ is not simple, we get that $R$ is the direct sum of a division ring and its opposite.

Finally, suppose that $R$ is simple, and not a division ring. If every non-zero symmetric element of $R$ is invertible, then we are done by the preceding theorem. If $s^* = s$, then $2s = s + s^*$ is invertible or 0; since not every $s \neq 0 \in S$ is invertible we see that char $R = 2$. Let $a^* = a \neq 0$ be non-invertible. If $T = \{x + x^* \mid x \in R\}$, then, since $aTa \subset T$ and the elements of $aTa$ are not invertible, $aTa = 0$. This gives $axa = ax^*a$ for all $x \in R$. Thus, if $x, y \in R$, $a(xay)a = a(xay)^*a = ay^*ax^*a = ayaxa$.

Since $R$ is a simple ring, has a unit element, and $axaya = ayaxa$ for all $x, y \in R$, where $a \neq 0$, by the corollary to Theorem 1.3.1, $R$ is isomorphic to $F_n$, the ring of $n \times n$ matrices over a field $F$. We claim that $n = 2$. We claim that it is not possible that $e^* = e$ for all minimal idempotents $e$. If this were the case, since $f = e + xe + exe$ is also a minimal idempotent, $f^* = f$. From $f^* = f$ we get that $ex^* + xe = e(x + x^*)e$ so is in $eTe$. Because $eTe \subset T$ and no element of $eTe$ is invertible, $eTe = 0$. Thus we get that $ex = x^*e$ for all $x \in R$, and so $Re = eR$ is a proper ideal of $R$, which is not possible since $R$ is simple.

Therefore $e + e^* \neq 0$ for some minimal idempotent $e$. But $e + e^*$ is then invertible, hence $R = eR + e^*R$ is the sum of two minimal right ideals, namely $eR$ and $e^*R$. This says that $n = 2$. Knowing that $x + x^* = 0$ must be invertible, it is then an exercise to show that the involution on $F_2$ is symplectic.

Example 3. Let $R$ be a ring with involution and let $S^2$ be the additive group of $R$ generated by all $ab$ where $a, b \in S$. It is easy to see that $S^2$ is a Lie ideal of $R$. For, suppose that $a, b \in S$ and $x \in R$; then $abx - xab = a(bx + x^*b) - (ax^* + xa)b$, so is in $S^2$.

Note that the proof shows a little more, namely, that if $a, b \in T$ then $abx - xab$ is, in fact, in $T^2$. Thus $T^2$ is a Lie ideal of $R$.

In connection with this Lie ideal, $S^2$, of $R$ there is an open question for simple rings. If $R$ is simple, we say that $*$ is of the first kind if $\alpha^* = \alpha$ for all $\alpha \in Z(R)$; otherwise we say that $*$ is of the second kind.

Question. If $R$ is simple and $*$ is of the first kind, is $S^2 = R$ if $\dim_Z R > 4$?

This question is equivalent to the following one concerning the skew elements $K$, namely, if $R$ is simple and $*$ is of the first kind, is $K = [K, K]$? For matrices the answer to these questions is yes.

We return to Osborn's theorem for a moment. In proving Theorem 2.1.8 we assumed the ring to be semi-prime. What happens if we allow $R$ to have nilpotent ideals ?

So, let $R$ be a ring with involution in which every non-zero symmetric element is invertible. We claim that the maximal nil ideal, $N$, of $R$ must be nilpotent. If $N = 0$ there is nothing to prove. If $N \neq 0$, then, since $N^* = N$, if $a \neq 0$ is in $N$ then $a + a^*$ is also in $N$. If $a + a^* \neq 0$, then by our assumption $a + a^*$ is invertible, which it is not, since it is nilpotent. Thus $a + a^* = 0$ for every $a \in N$, and so $a^* = -a$ for every $a \in N$. Since $N \cap S = 0$, we see that $N$ must be 2-torsion free. If $a \in N$ then from $a^* = -a$ we have $a^2 = (a^2)^* \in N \cap S = 0$, so $a^2 = 0$. Because $N$ is 2-torsion free we have that $N^3 = 0$. Since $2N \neq 0$, $2 = 1 + 1^* \neq 0$ is invertible in $R$. Thus $R/N$ is 2-torsion free.

If $a \in N$ then $a^* = -a$; if $x \in R$ then, since $xa \in N$, $(xa)^* = -xa$. Proceeding as in the proof of Theorem 2.1.8, we get that $[R, R]N = 0$, and so $\mathcal{C}N = 0$ where $\mathcal{C}$ is the commutator ideal of $R$. Now, $\mathcal{C}^* = \mathcal{C}$ and $\mathcal{C}$ annihilates $N \neq 0$, hence $\mathcal{C}$ can have no invertible elements, whence $\mathcal{C} \cap S = 0$. Proceeding as we did with $N$ leads us to $\mathcal{C}^2 = 0$ and so $\mathcal{C} \subset N$.

Hence $R/N$ is commutative. Because it is 2-torsion free, its non-zero symmetric elements are images of symmetric elements in $R$, and so are invertible in $R/N$. Invoking Theorem 2.1.7, $R/N$ is either a field $F$, or the direct sum of $F$ with itself. We have proved

THEOREM 2.1.9. Let $R$ be a ring with involution in which all $s \neq 0$ in $S$ are invertible in $R$. If $R$ has a non-zero nilpotent ideal, then it has a maximal nilpotent ideal $N$, $N^3 = 0$ and $R/N$ is (2-torsion free) either a field or the direct sum of a field and itself.

Example 4. Let $R$ be a ring with involution and let $K^2$ be the additive group generated by all $ab$, $a, b \in K$. Then $K^2$ is a Lie ideal of $R$. For, if $a, b \in K$ and $x \in R$ then $abx - xab = a(bx + x^*b) - (ax^* + xa)b$; since $bx + x^*b = bx - (bx)^*$ and $ax^* + xa = ax^* - (ax^*)^*$, we get that $abx - xab \in K^2$, hence $K^2$ is a Lie ideal of $R$.

Note that the same proof reveals that $K_0^2$ is a Lie ideal of $R$, where $K_0 = \{x - x^* \mid x \in R\}$ is the set of skew-traces of $R$.

Theorems 2.1.7, 2.1.8 and 2.1.9 have their analogs when the conditions are imposed on $K$ and $K_0$ rather than on $S$ and $T$. We shall do these later in this chapter. In the meantime we do the skew analog of Theorem 2.1.6 now.

THEOREM 2.1.10. Let $R$ be a simple ring with involution. If $\dim_Z R > 4$, then $\overline{K_0} = R$ where $\overline{K_0}$ is the subring of $R$ generated by the skew-traces $K_0$.

<u>Proof</u>. If char $R = 2$, then $K_0 = T$, and so $\overline{K_0}$ is a Lie ideal of R. As such, since $\dim_Z R > 4$, by Theorem 2.1.3 we have that either $\overline{K_0} \subset Z = Z(R)$, or $\overline{K_0} \supset [R, R]$. In this latter case, $\overline{K_0}$ contains the subring which is generated by $[R, R]$, which is easily seen to be $R$. So we may suppose that $\overline{K_0}$, and so $K$, is contained in $Z$.

If $x \in R$ then $x + x^* \in K_0 \subset Z$. Since $K_0 \neq 0$ (otherwise $R$ is commutative), there is a $u \in R$ such that $\alpha = u + u^* \neq 0$ is in $Z$. Now $\alpha x^* x = x^*(u + u^*)x$ so is in $K_0 \subset Z$; in consequence, $x^* x \in Z$. Thus $x$ satisfies $x^2 - (x + x^*)x + x^* x = 0$, a quadratic equation over $Z$, for every $x \in R$. This gives the contradiction that $\dim_Z R \leq 4$. So, if char $R = 2$ we are done. Hence we may assume that char $R \neq 2$, and so $K_0 = K$.

Since $K^2$ is a Lie ideal of $R$, the subring it generates, $\overline{K^2}$, is both a subring and Lie ideal of $R$. By Theorem 2.1.2, $\overline{K^2} = R$ or $K^2 \subset Z$. So, if $\overline{K} \neq R$ we must have that $K^2 \subset Z$. Our objective will be to show that $\dim_Z R \leq 4$ results from this.

Given $a, b \in K$ then $ab \in K^2 \subset Z$, hence since $Z$ is a field, $ab = 0$ or $a$ must be invertible in $R$. If $a$ is not invertible, we get $aK = 0$. If $s \in S$ then $sas \in K$ hence $asas = 0$. Given $x \in R$, $x = s + k$ where

$s = \frac{x+x^*}{2} \in S$, $k = \frac{x-x^*}{2} \in K$; therefore $(ax)^3 = a(s+k)a(s+k)a(s+k) = 0$, which is to say every element in $aR$ has cube 0. But his violates the fact that $R$ is simple, hence semi-prime, (by Lemma 2.1.1) unless $a = 0$. In short, $a \neq 0 \in K$ implies that $a$ is invertible in $R$.

If $0 \neq a \in K$ then $a^2 = \mu \neq 0 \in Z$, whence $a^{-1} = \mu^{-1}a$. Since $aK \subset Z$, we have that $K \subset Za^{-1} = Za$. If $s \in S$ commutes with $a$, since $sa$ must then be in $K$, $sa = \alpha a$ for some $\alpha \in Z$, leading us to $s = \alpha \in Z$. In other words, the only symmetrics commuting with $a$ are those in the center.

Given $s \in S$, $sa + as \in K$, hence $sa + as = \mu_1 a$, $\mu_1 \in Z \cap S$. Thus $(s - \frac{\mu_1}{2})a + a(s - \frac{\mu_1}{2}) = 0$. But then, $(s - \frac{\mu_1}{2})^2$ commutes with $a$; being symmetric, $(s - \frac{\mu_1}{2})^2$ must be in $Z$.

Given $x \in R$ then $x = s + k$ where $s \in S$ and $k \in K$, hence $x = s + \lambda a$ where $\lambda \in Z$. Since $sa + as = \mu a$, $\mu \in Z$, we have as above that $(s - \frac{\mu}{2})^2 \in Z$. Thus $(x - \frac{\mu}{2})^2 = \lambda(a(s - \frac{\mu}{2}) + (s - \frac{\mu}{2})a) + \lambda^2 a^2 = (s - \frac{\mu}{2})^2 + \lambda^2 a^2$ so is in $Z$ since both $a^2$ and $(s - \frac{\mu}{2})^2$ are in $Z$. Therefore $R$ is quadratic over $Z$, in consequence of which, $\dim_Z R \leq 4$, a contradiction.

Baxter [1, 3] showed that if a ring is generated by its skew elements then it is done so in a very particular way. We prove a generalization of his result here.

THEOREM 2.1.11. Let $R$ be a ring with involution in which $2R = R$. If $\overline{K} = R$ then $S = K \circ K$, hence $R = K + K \circ K$. In other words, given $s \in S$ then $s = \sum k_i^2 - \sum q_j^2$, where $k_i \in K$, $q_j \in K$.

Proof. If $a, b \in K$ then $ab - ba \in K$, $ab + ba \in K \circ K$ and so $2ab = (ab - ba) + (ab + ba) \in K + K \circ K$. Since $2R = R$ we get from this that $K^2 \subset K + K \circ K$.

We claim that $(K \circ K)K \subset K + K \circ K$. For let $a, c \in K$; then $a^2c + ca^2 \in K$, $a^2c - ca^2 = a(ac - ca) + (ac - ca)a \in K \circ K$, whence $2a^2c \in K + K \circ K$, and so $a^2K \subset K + K \circ K$. Linearize this on $a$, getting thereby $(K \circ K)K \subset K + K \circ K$.

Now $K^2 \subset K + K \circ K$, hence $K^3 = K^2K \subset K^2 + (K \circ K)K \subset K + K \circ K$. Continuing, we get $K^n \subset K + K \circ K$ for all $n$. But $\overline{K} = \sum K^n$ and, since $\overline{K} = R$, $R = \sum K^n \subset K + K \circ K$. Thus $R = K + K \circ K$. Since $2R = R$, we get from this, $S = K \circ K$.

The theorem has the following two interesting consequences.

COROLLARY 1. Let $R$ be a simple ring of characteristic not 2. Then, if $\dim_Z R > 4$, $S = K \circ K$.

Proof. If $\dim_Z R > 4$, then by Theorem 2.1.10, $\overline{K} = R$. Since char $R \neq 2$, Theorem 2.1.11 applies, and we have the corollary.

COROLLARY 2. If $R$ is simple, with $*$, of characteristic not 2, and if $K$, as a vector space over $Z \cap S$ is finite-dimensional, of dimension $n$, then $\dim_Z R \leq n^2 + n$ (or 4).

Proof. If $\dim_Z R \leq 4$ there is nothing to prove. If $\dim_Z R > 4$, by Corollary 1, $R = K + K \circ K$. If $v_1, \ldots, v_n$ is a basis of $K$ over $Z \cap S$, then $v_iv_j + v_jv_i$, $i, j = 1, 2, \ldots, n$ generate $K \circ K$ over $Z \cap S$. Thus $\dim_Z K \circ K \leq n^2$, hence $\dim_Z R = \dim_Z (K + K \circ K) \leq n^2 + n$.

We close this section with a discussion of the ideal structure of S, as a Jordan ring, in semi-prime rings. The result is

THEOREM 2.1.12. Let R be a semi-prime, 2-torsion free ring with involution. Suppose that $U \neq 0$ is a Jordan ideal of S. Then there exists an ideal $W \neq 0$ of R such that $U \supset \{w + w^* \mid w \in W\}$.

Proof. If $a \in U$, $s \in S$ then, since $2a^2 = aa + aa \in U$, $4sas = 2\{(as + sa)a + a(as + sa)\} - 2(a^2s + sa^2) \in U$. A similar argument shows that $4asa \in U$.

If $x \in R$ then $4(a^2x + x^*a^2) = 4\{a(ax + x^*a) + (ax + x^*a)a\} - 4a(x + x^*)a$ so, by the above, is in U. That is, $4(a^2x + x^*a^2) \in U$ for all $a \in U$, $x \in R$. Hence $2\{4(a^2x + x^*a^2)\}^2 \in U$; expanding this, we have $32\{a^2xa^2x + x^*a^2x^*a^2\} + 32a^2xx^*a^2 + 32x^*a^4x \in U$. By our results above, since $2a^2 \in U$, we have $32a^2xx^*a^2 \in U$ and $32\{a^2xa^2x + (xa^2x)^*a^2\} \in U$; the net result of all this is that $32x^*a^4x \in U$ for all $x \in R$, $a \in U$. Linearize this on x; this yields $32(x^*a^4y + y^*a^4x) \in U$. Let $W = 32Ra^4R$; W is an ideal of R and, moreover, if $w \in W$ then $w + w^* \in U$ follows from $32(x^*a^4y + y^*a^4x) = 32(x^*a^4y + (x^*a^4y)^*)$ being in U. So W does the trick, provided $W \neq 0$.

If $W = 0$, then since R is 2-torsion free, $Ra^4R = 0$ for all $a \in U$; since R is semi-prime, this would force $a^4 = 0$ for all $a \in U$. Since, for $a \in U$, $2a^2 \in U$, then for any $x \in R$, $4(a^2x + x^*a^2) \in U$ — as we saw earlier. Hence $(a^2x + x^*a^2)^4 = 0$. Multiply this from the right by $a^2$ and from the left by x; since $(a^2)^2 = a^4 = 0$, we get $(xa^2)^5 = 0$ for all $x \in R$. However, this forces the existence of a non-zero nilpotent ideal in $R$, unless $a^2 = 0$ for all $a \in U$. In consequence, $a^2 = 0$ for all $a \in U$. Linearize this on $a$; we get $ab + ba = 0$ for all $a, b \in U$. If $s \in S$, $b = as + sa \in U$ hence $0 = ab + ba = a(as + sa) + (as + sa)a = 2asa$, since $a^2 = 0$. But $R$ is 2-torsion free; hence $aSa = 0$ for all $a \in U$. If $k \in K$ then $kak \in S$, therefore, $akaka = 0$. If $x \in R$, then $2x = s + k$ where $s = x + x^*$, $k = x - x^*$; thus $(2ax)^3 = a(s + k)a(s + k)a(s+k) = 0$, and so, $(ax)^3 = 0$. This is not possible in a semi-prime ring. In short $a = 0$ follows. Since $U \neq 0$ we must have $W = 32Ra^4R \neq 0$ for some $a \in U$. As we have seen, $w + w^* \in U$ for all $w \in W$.

Note that $\{w + w^* \mid w \in W\} \neq 0$, otherwise $w = -w^*$ for all $w \in W$; this leads to $w^2 = 0$ and so $W^3 = 0$ in the 2-torsion free ring. Because $W \neq 0$ and $R$ is semi-prime, this too is not possible.

The theorem has a very interesting consequence, namely (TRT Theorem 2.6).

COROLLARY. Let $R$ be a simple ring with involution of characteristic not 2. Then $S$ is a simple Jordan ring.

## 2. Positive-definiteness Theorems

We begin with a lemma which we have used explicitly or implicitly several times so far. It is understood that R will be a ring with involution * throughout.

LEMMA 2.2.1. Let $U \neq 0$ be an ideal of R such that $U^* = U$. If $U \cap S = 0$ then $U^3 = 0$.

Proof. If $x \neq 0$ is in U then $x + x^* \in U \cap S = 0$, hence $x^* = -x$. Since $-x^2 = xx^* \in U \cap S = 0$, we have $x^2 = 0$ for all $x \in U$. Furthermore, since $U \cap S = 0$ and $x^* = -x$ for all $x \in U$, U must be 2-torsion free. But then, trivially, $U^3 = 0$.

We now prove the first result of this positive-definiteness type. It will assert that under certain conditions imposed on R, $xx^* \neq 0$ if $x \neq 0$. These types of results are very important for the purposes of reducing complicated situations to more specialized and simpler ones. We'll see how this reduction works later on. In the meantime we prove [4]

THEOREM 2.2.1. Let R be a prime ring in which no non-zero element of S is nilpotent. Then either

1. $xx^* \neq 0$ in R if $x \neq 0$, or

2. $S \subset Z(R)$ and R is an order in $F_2$, the ring of all $2 \times 2$ matrices over a field F (relative to the symplectic involution).

Proof. Suppose that $xx^* = 0$ for some $x \neq 0$ in R. We want to show that the second possibility stated in the theorem must then prevail.

Now, $x^*Sx \subset S$ and, from $xx^* = 0$, every element in $x^*Sx$ is nilpotent. Therefore, $x^*Sx = 0$; so, if $r \in R$ then $x^*(r + r^*)x = 0$, which is to say, $x^*rx = -x^*r^*x = -(x^*rx)^*$. Now, because R is prime, $k = x^*rx \neq 0$ for some $r \in R$; k is skew by the above, and $k^2 = 0$. Since k is nilpotent it cannot be symmetric by hypothesis, in consequence of which the <u>characteristic of</u> R <u>cannot be</u> 2.

Since $0 = -k^2 = -kk^*$, as at the beginning of this proof, we have $kSk = 0$. Now, if $s \in S$ then $sk - ks \in S$; but $(ks - sk)^2 = ksks + sksk - sk^2s - ks^2k = 0$ since $kSk = 0$ and $k^2 = 0$. Our hypothesis on S then tells us that $sk = ks$ for all $s \in S$, that is, k centralizes S and therefore centralizes $\overline{S}$. However, by Theorem 3.1.5, $S \subset Z$, or $\overline{S}$ contains a non-zero ideal $U \neq 0$ of R. In this second possibility, since k centralizes U, by Lemma 1.1.6, k must be in Z(R). But Z(R) has no nilpotent elements. The upshot of this is that $S \subset Z(R)$.

We claim that this will force R to be an order in $F_2$. Localize R at S to get the ring $R_S = \{r/z \mid r \in R, z \neq 0 \in S\}$. $R_S$ is clearly prime with an involution. Its non-zero symmetric elements are all invertible. Moreover, $R_S$ has a non-zero nilpotent element, namely $k/z$. Theorem 2.1.7 applies here; its result then allows as the only possible answer here that $R_S = F_2$. But, since every element in $R_S$ is of the form $rz^{-1}$ where $r \in R$, $z \neq 0 \in S$, if r is regular in R it must be regular in $R_S = F_2$, so is invertible in $R_S$. Thus we conclude that R is an order in $F_2$, completing the proof.

There are many prime rings of operators on a Hilbert space which satisfy the hypothesis of the preceding theorem. They are quite disparate; the one common property they do share is that $xx^* \neq 0$ if $x \neq 0$. So the theorem does seem to give about as good an answer as one could expect.

We shall soon prove a skew analog of Theorem 2.2.1. Before doing so we digress to prove an interesting result due to Lanski [6]. Our first step towards this theorem of Lanski, which deals with semi-prime rings, is to dispose of the prime case.

LEMMA 2.2.2. Let $R$ be prime and suppose that $a \neq 0$, $b \neq 0$ in $S$ implies that $ab \neq 0$. Then either

1. $R$ is a domain, or

2. $S \subset Z(R)$ and $R$ is an order in $F_2$ for some field $F$.

Proof. Suppose that $R$ is not an order in $F_2$. Since $S$ has no non-zero nilpotent elements, by Theorem 2.2.1, if $xx^* = 0$ then $x = 0$.

Suppose now that $xy = 0$ for $x, y \in R$. Hence $(x^*x)(yy^*) = 0$; because $x^*x, yy^*$ are both in $S$, by our hypothesis $x^*x = 0$ — in which case, $x = 0$ — or $yy^* = 0$ — in which case, $y = 0$. Thus $R$ must be a domain.

We now can obtain the full result proved by Lanski.

THEOREM 2.2.2. Let $R$ be a semi-prime ring in which $a \neq 0$, $b \neq 0$ in $S$ implies that $ab \neq 0$. Then

1. $R$ is a domain, or

2. $S \subset Z(R)$ and $R$ is an order in $F_2$, $F$ a field, or

3. $R$ is a subdirect sum of a domain and its opposite, with the involution the exchange involution.

Proof. If $R$ is prime, we saw in Lemma 2.2.2 that either conclusion 1 or 2 must hold. So we may suppose that $R$ is not prime.

So we suppose there are ideals $A \neq 0$, $B \neq 0$ of $R$ such that $AB = 0$. Hence $A^*AB = 0$; this gives that $B^*A^*A = 0$, and so, by the semi-primeness of $R$, $A^*AB^* = 0$. Thus we have that $(A^*A)(B+B^*) = 0$. By Lemma 2.2.1, there is a non-zero symmetric element $b$ in $B+B^*$; since $(A^*A)b = 0$, $A^*A$ cannot have a non-zero symmetric element. By Lemma 2.2.1, again, we get that $A^*A = 0$. Therefore, whenever $AB = 0$, $A, B$ non-zero ideals of $R$, then $A^*A = 0$, and so, $A \cap A^* = 0$.

Let $M = \{x \in R \mid xA = 0\}$; $M \supset A^*$ is an ideal of $R$. Because $MA = 0$, from the above, $M^*M = 0$, and so, since $R$ is semi-prime, $MM^* = 0$, $M \cap M^* = 0$. If $Mx = 0$ then $A^*x = 0$ since $M \supset A^*$, hence $x^*A = 0$ and so $x^* \in M$. What this says is that $M^*$ is precisely the annihilator in $R$ of $M$.

Note, too, that $M$ has no non-zero nilpotent elements. For if $m \neq 0 \in M$ and $m^2 = 0$ then $m + m^* \neq 0$ and $(m+m^*)^2 = m^2 + (m^*)^2 = 0$ since $MM^* = M^*M = 0$; this contradicts the basic hypothesis on $S$. We claim more, namely, that $M$ is a domain. For, if $u, v \in M$ and $uv = 0$ then, since $(vu)^2 = 0$, we have that $vu = 0$. But $(u+u^*)(v+v^*) = uv + u^*v^* = uv + (vu)^* = 0$, forcing $u+u^* = 0$ or $v+v^* = 0$. If $u+u^* = 0$ then $u = -u^* \in M \cap M^* = 0$, which implies that $u = 0$. Similarly, if $v+v^* = 0$ then $v = 0$. Thus $M$ is, indeed, a domain.

We assert that $R/M^*$ is a domain. For, suppose that $xy \in M^*$. If $u, v \in M$ then $ux, yv$ are both in $M$ and $(ux)(yv) = u(xy)v$ is in $M \cap M^* = 0$. Consequently $ux = 0$ or $yv = 0$ since $M$ is a domain. This easily gives that $Mx = 0$ or $yM = 0$; but if $Mx = 0$ then $xM = 0$. Hence, $xy \in M^*$ implies that $xM = 0$ or $yM = 0$. However, as we saw, $M^*$ is <u>precisely</u> the annihilator of $M$. Consequently, $xy \in M^*$ implies that $xM = 0$ or $yM = 0$, which implies that $x \in M^*$ or $y \in M^*$. Thus

$R/M^*$ is a domain; similarly $R/M$ is a domain. Because $M \cap M^* = 0$, $R$ is a subdirect sum of $R/M$ and $R/M^*$.

To verify that the involution on R as embedded in $R/M \oplus R/M^*$ is the exchange involution is trivial.

Clearly, Lanski's theorem generalizes that of Osborn. For Osborn's result we were able to get information even when the ring has nilpotent ideals (Theorem 2.1.9). The same is also almost true for the result of Lanski; however we must impose a slightly more exigent hypothesis. The reason for this is that if we merely insist that the elements of S are not zero-divisors on S, they may very well be zero divisors in R, when R is not semi-prime. It leads to a related question: if a given $a \in S$ is not a zero divisor on S, when can we be sure it is not a zero-divisor in R? We shall go into this later. Now we go into the non-semi-prime version of Lanski's theorem.

THEOREM 2.2.3. Let R have a non-zero nilpotent ideal, and suppose that all non-zero $x+x^*$ are not zero-divisors in R. Then, if N is the maximal nil ideal of R, $R/N$ is commutative.

Proof. Since $N^* = N$ and no non-zero $x + x^*$ is nilpotent, we get — as we have several times — that if $a \in N$ then $a^* = -a$, and $ra = ar^*$ for all $r \in R$. As before, this implies that $[R, R]N = 0$ and so $\mathfrak{C}N = 0$, where $\mathfrak{C}$ is the commutator ideal of R. Since every element in $\mathfrak{C}$ is a zero-divisor — annihilating $N \neq 0$ — as we did for N, we get that $[R, R]C = 0$ and so $\mathfrak{C}^2 = 0$. Thus $\mathfrak{C} \subset N$ and $R/N$ is commutative.

We now propose to prove a skew version of Theorem 2.2.1; our hypothesis is slightly weaker in that we impose it only on the skew-traces $K_0$, instead of on all of $K$.

THEOREM 2.2.4. Let $R$ be a prime ring in which no non-zero element in $K_0 = \{x - x^* \mid x \in R\}$ is nilpotent. Then either

1. $xx^* \neq 0$ in $R$ if $x \neq 0$, or

2. $R$ is an order in $F_2$, where $F$ is a field.

Proof. If $K_0 = 0$ then every element in $R$ must be symmetric. Hence $R$ must be commutative. In that case, since $R$ is prime, $R$ is a commutative integral domain, hence we certainly have $xx^* \neq 0$ for $x \neq 0$ in $R$. Therefore we may assume that $K_0 \neq 0$. In fact, we may also suppose that $R$ is not commutative, for otherwise $R$ is a commutative integral domain.

Suppose that $xx^* = 0$ for some $x \neq 0$ in $R$. Since $x^*K_0x \subset K_0$ and consists of nilpotent elements, $x^*K_0x = 0$ , and so, $x^*rx = x^*r^*x = (x^*rx)^*$ for all $r \in R$. But, because $R$ is prime, there is an $r \in R$ with $s = x^*rx \neq 0$; $s \in S$, $s \neq 0$ but $s^2 = 0$. Therefore, $sK_0s = 0$. So, if $k \in K_0$, then $sk + ks \in K_0$; since $(sk+ks)^2 = sk^2s$ and so $(sk+ks)^4 = (sk^2s)^2 = 0$, by our hypothesis on $K_0$, $sk+ks = 0$. Therefore $s$ centralizes $K_0^2$.

However, $K_0^2$ is a Lie ideal of $R$, so the subring it generates, $\overline{K_0^2}$ , either contains a non-zero ideal of $R$ or $\overline{K_0^2}$ is commutative and the square of any element in $\overline{K_0^2}$ is in $Z = Z(R)$. Now $s \in S$ is non-central because it is nilpotent, so $s$ cannot centralize a non-zero ideal of R. In consequence, $w \in \overline{K_0^2}$ implies that $w^2 \in Z$. In particular, if $u \neq 0 \in K_0$, then $u^4 \neq 0$ must be in $Z$. In particular, $Z \neq 0$.

Localize R at $Z^+ = Z \cap S$; if we call this localization Q, then * extends to Q. Moreover, if $K_1$ is the set of skew-traces of Q then $K_1^2 \subset Z_1 = Z(Q)$, and if $k \neq 0 \in K_1$ then $k^2 \neq 0 \in Z_1$. Thus k is invertible in Q. We claim that Q is simple. For, if $V \neq 0$, V is an ideal of Q, then $W = VV^* \neq 0 \subset V$ is a *-ideal of Q. It cannot contain any non-zero skew-traces, for these are invertible. Hence W must consist of symmetric elements, so, as a ring, W is commutative. Since Q is prime and contains a non-zero commutative ideal, by the corollary to Lemma 1.1.5 we get that Q is commutative, a contradiction. In this way we see that Q is simple.

If u, v are two non-zero skew-traces in Q then $uv \neq 0$ is in $Z_1$, and since $Z_1$ is a field, $u = \alpha v$, where $\alpha \in Z_1$, follows. But then, by Theorem 2.1.10, $\dim_{Z_1} Q = 4$. Since Q has zero-divisors (after all, $s^2 = 0$ and $s \neq 0$), Q must be the ring of all $2 \times 2$ matrices over a field F. Since R is an order in Q, the proof of the theorem is complete.

As a corollary to this theorem, we can sharpenTheorem 2.2.1.

COROLLARY. Let R be a prime ring in which no non-zero $x+x^*$ is nilpotent. Then, either

1. $xx^* \neq 0$ if $x \neq 0$, or

2. R is an order in $F_2$, where F is a field.

<u>Proof.</u> If char R = 2, the corollary is merely Theorem 2.2.4. If char $R \neq 2$ the condition on $x+x^*$ tells us that for $0 \neq s \in S$, 2s is

not nilpotent, hence $s$ is not nilpotent. By Theorem 2.2.1, the corollary follows.

3. A Skew Version of Osborn's Theorem

Recall that if $R$ is a ring with involution $*$ then $K_0 = \{x - x^* \mid x \in R\}$ is the set of skew-traces of $R$. In Theorems 2.1.7 and 2.1.8 we classified semi-prime rings whose non-zero symmetric elements, or traces, are invertible in $R$. In Theorem 2.1.9 we got a rather tight description of the structure of $R$, even when it is not semi-prime, when we know that its non-zero symmetric elements are invertible.

In this section we propose to prove skew analogs of the theorems cited above. These results come from the joint paper [5] by Herstein and Montgomery.

THEOREM 2.3.1. Let $R$ be a non-commutative semi-prime ring in which every non-zero element of $K_0$ is invertible in $R$. Then $R$ is

1. a division ring, or

2. the direct sum of a division ring and its opposite, relative to the exchange involution, or

3. the ring $F_2$ of $2 \times 2$ matrices over a field $F$.

Proof. We claim that $R$ has no ideal $A \neq 0$, $R$ such that $A^* = A$. For, if $A$ is a non-trivial $*$-ideal of $R$, since $A$ has no invertible elements, $a = a^*$ for all $a \in A$. But then $A$ is commutative, so by the corollary to Lemma 1.1.5, $A \subset Z$. This implies that $A[R, R] = 0$, hence $A\mathfrak{C} = 0$ where $\mathfrak{C}$ is the commutator ideal of $R$. But $\mathfrak{C}^* = \mathfrak{C}$ and since

$A\mathfrak{C} = 0$, $A \neq 0$, we must have that $\mathfrak{C}$ has no invertible elements; thus, as above, $\mathfrak{C}[R, R] = 0$. This gives $\mathfrak{C}^2 = 0$ and so $\mathfrak{C} = 0$; however, $\mathfrak{C} = 0$ implies that R is commutative, contrary to assumption. In other words, R must be $*$-simple.

If R is not simple, let $I \neq 0, R$ be an ideal of R. Then, by the remarks above, $I + I^* = R$ and $I \cap I^* = 0$, that is, R is the direct sum of I and $I^*$. The condition "$x - x^* \neq 0$ implies $x - x^*$ invertible" then trivially implies that I is a division ring. So, if R is not simple, it is the direct sum of a division ring and its opposite.

Suppose, then, that R is simple. If every $s \neq 0 \in S$ is invertible in R, then we have the result by applying Theorem 2.1.7. Suppose then, that $s \neq 0$ in S is not invertible in R. Hence no element of $sK_0s$ is invertible; since $sK_0s \subset K_0$, by our hypothesis on $K_0$ we must have that $sK_0s = 0$. As we have seen several times, this implies that $sxsys = sysxs$ for all $x, y \in R$. By the corollary to Theorem 1.3.1 we get that $R = F_n$, the ring of $n \times n$ matrices over a field .

Every $x \in F_n$ is a sum of elements of rank 1 as matrices, and since $x^* \neq x$ for some $x \in F_n$, we have that $u^* \neq u$ for some u of rank 1. But then $u^*$ is of rank 1 and $u - u^* \neq 0$ is of rank at most 2. However, by assumption, $u - u^*$ is invertible, in consequence of which, $n = \text{rank}(u - u^*) \leq 2$. Thus indeed $R = F_2$ as claimed in the theorem.

As in our previous theorems of this type, when there is a nilpotent ideal present we can get rather precise information about $R/N$, where N is the maximal nil ideal of R.

THEOREM 2.3.2. Suppose that $N \neq 0$ is the maximal nil ideal of $R$, and that all $x - x^* \neq 0$ are invertible in $R$. Then, if $R$ is not commutative,

1. $R/N$ is commutative
2. $N^2 = 0$.

Proof. As we saw in the proof of the preceding theorem, if $A \neq 0, R$ is a $*$-ideal of $R$ then $A\mathfrak{C} = 0$, where $\mathfrak{C}$ is the commutator ideal of $R$. Since $R$ is not commutative, $K_0 \neq 0$ hence $R$ has invertible elements. Thus $N \neq R$. By assumption, $N \neq 0$. But $N^* = N$, consequently $N\mathfrak{C} = 0$. Since $\mathfrak{C}$ is a $*$-ideal, $\mathfrak{C} \neq 0$ since $R$ is not commutative, and $\mathfrak{C} \neq R$ since $N\mathfrak{C} = 0$, we must have $\mathfrak{C}\mathfrak{C} = \mathfrak{C}^2 = 0$. So $\mathfrak{C} \subset N$ and $R/N$ is commutative.

If $a \in N$ then, because $N \cap K_0 = 0$, $a = a^*$; if $x \in R$ then, because $xa \in N$, $xa = (xa)^* = ax^*$. If $b \in N$ then $xab = ax^*b = abx$, thus $ab \in Z$. But $ab = ba$ is symmetric, hence $abK_0 \subset K_0$. No element in $abK_0$ is invertible, so $abK_0 = 0$ and so $ab = 0$. Thus $N^2 = 0$.

We should like to extend Theorem 3.1.1 to a less restrictive situation. Instead of insisting that every non-zero element of $S$, or $T$, or $K_0$ be invertible, we want to add another option, namely, that such an element be either invertible or nilpotent. This situation arises naturally in the theory of Jordan algebras. However, we are immediately faced with a difficulty: what if every element of $S$, say, is nilpotent ?

So, suppose that $s \in S$ implies that $s$ is nilpotent. A reasonable conjecture would be that $R$ itself must be nil. If this were not so, then

$N \neq R$, where $N$ is the maximal nil ideal of $R$. Since $N^* = N$, $*$ induces an involution on $R/N$; if $\bar{x}^* = \bar{x}$ for $\bar{x} \in R/N$ then $\bar{x}^2 = \bar{x}\bar{x}^* = \overline{xx^*}$, so $\bar{x}^2$ is the image of a symmetric element in $R$, so is nilpotent. Thus, if $N \neq R$, we would obtain (using $R/N$) a ring, without nil ideals, in which every symmetric element is nilpotent. If $0 \neq s \in S$, $s^2 = 0$ then, since $sx + x^*s \in S$ for every $x \in R$, $(sx + x^*s)^n = 0$. Thus $0 = x(sx + x^*s)^n s = (xs)^{n+1}$. Hence $Rs$ is a nil left ideal of $R$. Here we run up against the Kothe conjecture. If the Kothe conjecture were true, $R$ would have a non-zero nil ideal contrary to hypothesis. Thus, in the presence of the truth of the Kothe conjecture we would have that $R$ is nil if $S$ is.

At any rate, this is the content of a question due to McCrimmon [8]:

<u>McCrimmon Conjecture.</u> If $S$ is nil then $R$ is nil.

We just saw that the Kothe conjecture implies that of McCrimmon. One might wonder if the converse implication also holds true; that is, does the McCrimmon conjecture imply that of Kothe?

At this state, since we are faced with this open question, to avoid it we will often impose the condition of semi-simplicity on the rings under discussion.

An argument like the one given above shows

LEMMA 2.3.1. If $K_0 \neq 0$ and every element in $K_0$ is nilpotent, then $R$ has a non-zero nil left ideal and $J(R) \neq 0$. Similarly, if $T \neq 0$ and every element in $T$ is nilpotent, then $R$ has a non-zero nil left ideal and $J(R) \neq 0$.

Proof. Let $k \neq 0 \in K_0$; then $k^n = 0$, $k^{n-1} \neq 0$ for some $n$. If $n-1$ is odd, then $a = k^{n-1}$ is in $K$; if $n-1$ is even, $a = k^{n-1}$ is in $S$. Now, $a^2 = 0$, and if $x \in R$ then $ax - x^*a \in K_0$ if $a \in S$ (i.e., if $n-1$ is even) and if $a \in K$, $ax + x^*a \in K_0$ (i.e., if $n-1$ is odd). Thus, in either case, $(ax \pm x^*a)^n = 0$, hence $0 = x(ax \pm x^*a)^n a = (xa)^{n+1}$. Thus $0 \neq Ra$ is a nil left ideal of $R$, hence is in $J(R)$, in consequence of which, $J(R) \neq 0$. A similar argument works for $T$.

The next result gives us a highly useful device for passing from simple and primitive rings to other simple and primitive rings, which are constructed in an intrinsic manner from the ring itself. The result is a prototype of a general class of theorems that can be proved in this vein.

LEMMA 2.3.2. Let $R$ be a primitive ring, $a, b \in R$ such that $ba \neq 0$. If $M = J(aRb)$ then $aRb/M$ is primitive, and $M$ satisfies $bMa = 0$ and $M^3 = 0$. Furthermore, if $R$ is simple then $M$ is the unique maximal ideal of $aRb$ and $aRb/M$ is simple.

Proof. Let $V$ be a faithful, irreducible $R$-module, and let $W = Vb$. Thus $W \neq 0$ is an $aRb$-module. If $v \in V$ is such that $vba \neq 0$ then $vbaRb = Vb = W$, that is, $aRb$ acts transitively on $W/W_0$ where $W_0 = \{w \in W \mid wa = 0\}$. Therefore $W/W_0$ is an irreducible $aRb$-module. If $M = J(aRb)$ then $M$ annihilates $W/W_0$, that is, $WM \subset W_0$ hence $WMa \subset W_0a = 0$; this translates into $VbMa = 0$, and so $bMa = 0$. Thus $M^3 \subset aRbMaRb = 0$.

If $x \in aRb$ annihilates $W/W_0$ then $Wx \subset W_0$ and so $Vb\,xa = 0$, giving us $bxa = 0$. In other words, the annihilator, $A$, of $W/W_0$ in aRb satisfies $A^3 = 0$ as above, so $A \subset M = J(aRb)$. Hence $aRb/M$ acts faithfully on $W/W_0$, and so $W/W_0$ is a faithful irreducible $\frac{aRb}{M}$-module. Thus $aRb/M$ is primitive.

If, in addition, R is simple, and $U \neq 0$ is an ideal of aRb then $aRbVaRb \subset U$. If $bUa \neq 0$ then $RbVaR = R$ whence $U = aRb$ follows. So every proper ideal, U, of aRb satisfies $bUa = 0$, and so $U^3 \subset aRbUaRb = 0$. This puts U in M; hence M is the unique maximal ideal of aRb.

For simple rings we are now able to generalize Theorem 2.3.1.

THEOREM 2.3.3. Let R be a simple, non-radical ring in which every element of $K_0$ (or, in which every element of T) is nilpotent or invertible. Then either R is a division ring or the ring of $2 \times 2$ matrices matrices over a field.

Proof. We give the argument for $K_0$; a very similar argument works for T. Suppose that R is not a division ring. Let $x \in R$ be not invertible in R. We claim that $x^*x = 0$. If not, by Lemma 2.3.2, $A = \frac{xRx^*}{M}$ is a simple, non-radical ring, where M is the unique maximal ideal of $xRx^*$. Every element in A of the form $\bar{u} - \bar{u}^* = \bar{u} - \overline{u^*} = \overline{u - u^*}$ is the image of $u - u^*$ in $xRx^* \cap K_0$. But no element in $xRx^*$ can be invertible in R, so all the elements of $xRx^* \cap K_0$ are nilpotent. If A is not commutative, by Lemma 2.3.1, $J(A) \neq 0$, a contradiction. So $\bar{u} = \bar{u}^*$ for every $\bar{u} \in A$, hence $u - u^* \in M$ for every $u \in xRx^*$. But, by Lemma 2.3.2, $x^*Mx = 0$; therefore $x^*ux = x^*u^*x$ for all $u \in xRx^*$. If $b = x^*x \neq 0$, this says that $brb = br^*b$ for all $r \in R$. The last part of the proof of Theorem 2.3.1 can be slightly modified to show that $R = F_2$, for some field $F$, and we would be done.

Hence we may assume that if $x \in R$ is not invertible, then $x^*x = 0$. In particular, if $a \in K_0$ is not invertible, then $0 = a^*a = -a^2$, hence $a^2 = 0$. If $x \in R$ is not invertible, then $x^*x = 0$, and so certainly $ax^*xa = 0$. On the other hand, if $x \in R$ is invertible, then $xa$ is not invertible, and so $(xa)^*xa = 0$; this gives $ax^*xa = 0$. In short, $ax^*xa = 0$ for all $x \in R$. Since $1 \in R$ then $a(x+1)^*(x+1)a = 0$ gives us that $a(x+x^*)a = 0$, and so $axa = -ax^*a$ for all $x \in R$. Thus $a(xay)a = -a(xay)^*a = ay^*ax^*a = ayaxa$. By the corollary to Theorem 1.3.1 we get that $R = F_n$ where $F$ is a field.

If $n > 2$, let $e^2 = e$ be of rank 1. Since $e$ and $e^*$ are not invertible, $ee^* = 0$ and $e^*e = 0$. Also, $e - e^*$ is of rank at most 2, so cannot be invertible; hence $(e - e^*)^2 = 0$. This gives $e + e^* = 0$, and so $0 = e(e+e^*) = e^2 = e$, a contradiction. Thus $n = 2$.

We are now able both to generalize Osborn's theorem from S to T, and also to prove a companion skew theorem. This is

THEOREM 2.3.4. Let $R$ be a ring, $J(R)$ its radical. If

1. every element in $T$ is nilpotent or invertible, or
2. every element in $K_0$ is nilpotent or invertible,

then $R/J(R)$ must be a division ring, the direct sum of a division ring and its opposite, the $2 \times 2$ matrices over a field, or a commutative ring with trivial involution.

Proof. Since $J(R)^* = J(R)$, it is clear that our hypothesis on $T$ or $K_0$ is inherited by $R/J(R)$. So we may assume that $R$ is semi-simple.

If $K_0 = 0$ then $R$ is commutative with trivial involution. If $T = 0$, then $x = -x^*$ for all $x \in R$, giving us $2x^2 = 0$ for all $x \in R$. But then $2R$ is a nil ideal of $R$, so $2R = 0$. Thus char $R = 2$, $T = K_0 = 0$, and $R$ must be commutative.

So we may assume that $T \neq 0$, $K_0 \neq 0$. In addition, by Lemma 2.3.1, we may assume that there are non-nilpotent elements in T and in $K_0$. Thus, according to hypotheses 1 or 2 there is an element $a \neq 0$ in T or $K_0$, accordingly, which is invertible.

We claim that R has no proper *-ideals. For if $I^* = I \neq R, 0$ then either $T \cap I = 0$ or $I \cap K_0 = 0$, according to which hypothesis we assume. If $I \cap K_0 = 0$, I is a commutative ideal in R, so $I \subset Z(R)$, and $I\mathcal{C} = 0$ where $\mathcal{C}$ is the commutator ideal of R. Since $\mathcal{C}^* = \mathcal{C} \neq 0$ and $I\mathcal{C} = 0$, $I \neq 0$, we have $\mathcal{C} \neq R$. If $\mathcal{C} \cap K_0 = 0$ then $\mathcal{C}^2 = 0$ as above. So $\mathcal{C} \cap K_0 \neq 0$. But no element of $\mathcal{C} \cap K_0$ can be invertible, so must be nilpotent. By Lemma 2.3.1, $J(\mathcal{C}) \neq 0$; but $J(\mathcal{C}) \subset J(R) = 0$, a contradiction. So we see that R is *-simple under hypothesis 2. A similar argument works under hypothesis 1. Thus in all situations, R is *-simple.

If R is simple, the result follows from Theorem 2.3.3 directly. So suppose that R is not simple; let I be a maximal ideal of R. By the *-simplicity of R, $R = I \oplus I^*$ and, since $I^* \approx R/I$, $I^*$ is simple; this tells us that I is simple. Because R is semi-simple, $I \neq J(I)$ otherwise $I \subset J(R)$; hence I is a simple, non-radical ring. Our hypothesis on T or $K_0$ implies that every element of I is nilpotent or invertible. However it is an easy exercise to show that in an associative ring in which every element is nilpotent or invertible the

the nilpotent elements form an ideal. Because I is simple we get that it has no nilpotent elements, hence, by our hypothesis, I must be a division ring. This completes the proof of the theorem.

## 4. Regular Skew Elements

In the last section we proved a skew analog of Osborn's theorem. In this section we shall prove an analog, for the skew elements, of Lanski's theorem (Theorem 2.2.2). Before getting to this we prove a general result which allows us to extend an involution from a prime ring to its central closure.

LEMMA 2.4.1. Let $R$ be a prime ring with involution $*$, and let $A = RC$ be the central closure of $R$. Then we can extend $*$ to an involution on $A$.

Proof. Let $\lambda \neq 0$ be in $C$; thus $\lambda = [f, U]$, where we can take $U = U^* \neq 0$, an ideal of $R$. Define $\lambda^* = [g, U]$ where $g(u) = (f(u^*))^*$. Since $\lambda \in C$, $f$ is a bi-module map of $U$ into $R$, from which we easily see that $g$ is a bi-module map of $U$ into $R$, hence $\lambda^* \in C$ follows. That $*$ defines an automomorphism of period 2 on C is trivial to verify. In fact it is also easy to see that $(r\lambda)^* = r^*\lambda^*$ for $r \in R$, $\lambda \in C$, if $r\lambda \in R$.

We define $*$ on $A = RC$ by $(\sum r_i c_i)^* = \sum r_i^* c_i^*$, $r_i \in R$, $c_i \in C$. We now show that this is well-defined. If $\sum r_i c_i = 0$, there is an ideal $0 \neq U = U^* \subset R$ such that $c_i U \subset R$ and $c_i^* U \subset R$ for all the $c_i$ we are using. If $u \in U$, then $\sum r_i(c_i u) = 0$, hence $(\sum r_i(c_i u))^* = 0$; because $c_i u \in R$, we get $\sum c_i^* u^* r_i^* = 0$, and so $u^* \sum r_i^* c_i^* = 0$ for all $u \in U$.

This gives, by primeness, that $\sum r_i^* c_i^* = 0$. Hence the $*$ defined on A is well-defined; that it is an involution extending that of R is immediate now.

We now prove the skew version of Lanski's theorem.

THEOREM 2.4.1. Let R be a non-commutative semi-prime ring in which all the non-zero elements of $K_0$ are regular in R. Then R is a domain, a subdirect sum of a domain and its opposite, or an order in $F_2$ for some field F.

Proof. We first handle the case in which R is prime; so suppose that R is a prime ring satisfying our hypothesis on $K_0$.

If every $s \neq 0 \in S$ were regular, by Lanski's theorem (Theorem 2.2.2) we would have that R is a domain or an order in $F_2$, and so we would be done. Thus we may assume that some $s \neq 0$ in S is a zero-divisor. But $sK_0s \subset K_0$ consists of zero-divisors, hence $sK_0s = 0$. This gives us $sxs = sx^*s$, and so $sxsys = sysxs$ for all $x, y \in R$.

Let $A = RC$ be the central closure of R; then A must also satisfy the G.P.I. $p(x, y) = sxsys - sysxs$. By Martindale's theorem, Theorem 1.3.2, A is a primitive ring having a minimal right ideal. But by Lemma 2.4.1, A itself is a ring with involution, hence by Kaplansky's theorem (Theorem 1.2.2) A contains a unique minimal ideal $M \neq 0$ consisting of all continuous linear transforrmations of finite range on a suitable vector space V. Since $M^*$ is an ideal of A, we get $M^* = M$.

Let $M_1 = M \cap R$; $M_1 \neq 0$ is an ideal of R, and $M_1^* = M_1$. If $x^* = x$ for all $x \in M_1$ then $M_1 \neq 0$ is a commutative ideal of R, and so we would get that R is commutative, and so an integral domain.

Thus $x - x^* \neq 0$ for some $x \in M_1$. But then $x - x^*$ is regular in R, hence is regular in A. However, since $x - x^*$ is in M, $x - x^*$ must be of finite range. This fact, together with the regularity of $x - x^*$ forces V to be finite-dimensional, in consequence of which, A is a simple artinian ring. But then every non-zero element in $K_0$, being regular in R, and so in A, must be invertible in A.

If $t = a - a^* \neq 0$ is a skew-trace in A, then t must be regular, hence invertible, in A. If not, there is an ideal $U \subset R$ such that all of the sets $aU$, $a^*U$, $aUa^*$ and $a^*Ua$ are in R. If $u \in U$, then $utu^* = uau^* - ua^*u$ must be in $U \cap K_0$. Since t is not regular, t is not invertible in A, therefore $utu^*$ is not invertible in A. Consequently $utu^*$ cannot be regular in A, and so, not in R. By our hypothesis on $K_0$, $utu^* = 0$. Also, by our choice of U, $t(u - u^*)t$ is in $K_0$ and cannot be regular, since it is not invertible. Thus $tut = tu^*t$. Therefore $0 = utu^*t = utut$. In short, $Ut \subset R$ is a nil ideal in which every element has square 0. This is not possible in a prime ring. With this contradiction we have that all the non-zero skew-traces in A must be regular, hence invertible in A. By Theorem 2.3.1 we deduce that either A is a division ring or $F_2$. Since R is an order in A, R must be a domain, or an order in $F_2$. This finishes the prime case.

Suppose, then, that R is semi-prime but not prime. Therefore $AB = 0$ for some ideals $A \neq 0$ and $B \neq 0$ of R. If $E = A \cap A^*$ and if $x \in E$ then $x - x^* \in E$ and $(x - x^*)B = 0$. By our hypothesis on $K_0$, $x = x^*$ for all $x \in E$. Thus E is a commutative ideal of R; by the corollary to Lemma 1.1.5, $E \subset Z(R)$, the center of R. Since R is not commutative, $K_0 \neq 0$. If $c \in E$ then $c^* = c \in Z(R)$, whence, if $k \neq 0 \in K_0$, then $ck \in K_0$. But $(ck)B = kcB = 0$; thus the element $ck \in K_0$ cannot be regular. Hence $ck = 0$. But $k \neq 0$ in $K_0$ is regular, from which we arrive at $c = 0$. In other words, $E = 0$, and so $A \cap A^* = 0$, and $AA^* = A^*A = 0$.

By Lanski's theorem there is an $s \neq 0$ in $S$ which is a zero-divisor, otherwise we are done. If $a \in A$, then $s(a - a^*)s \in K_0$ is a zero-divisor, therefore $s(a - a^*)s = 0$. This gives $sas = sa^*s \in A \cap A^* = 0$. Hence $sas = 0$ and so $(sA)^2 = 0$. Because $R$ is semi-prime, we get $sA = 0$. Similarly we get $sA^* = 0$. Therefore $s(a - a^*) = 0$ for all $a \in A$. Since $A \neq 0$, and as we saw above for $E$, not every $a \in A$ can satisfy $a = a^*$, we end up with $s = 0$. With this contradiction the proof of Theorem 2.4.1 is complete.

## 5. Some Theorems of Montgomery

We shall prove here a general theorem, proved by Montgomery [9], of what might be described as Faith-Utumi type, which holds in prime rings with involution. From this result we shall obtain a series of consequences. One of these describes prime rings with involution in which $xx^* = 0$ implies $x^*x = 0$. Another gives the description of prime, and semi-prime, rings whose traces, or whose skew traces, are regular or nilpotent. This last result finds application in the problem of the existence of a ring of quotients for quadratic Jordan algebras satisfying an Ore condition [10].

Of course the results of the preceding section, where skew traces were assumed to be regular, come out as special cases of these theorems. We developed the material as we did in order that the reader be exposed to a variety of the techniques that are used in this area of research.

In many of the theorems we have considered, where we have imposed some sort of "decency" condition on S, T, K or $K_0$ in a semi-prime ring R, R turned out to be a domain, a subdirect sum of a domain and its opposite, or an order in $F_2$, where F is a field. We view this trichotomy as a desirable sort of answer for the rings we are studying.

In looking at rings, even prime ones, in which every symmetric element, say, is regular or nilpotent we are forced with the possibility that all the symmetric elements are nilpotent. This puts us squarely into the framework of the open question raised by McCrimmon, namely, "must R then be nil ?". This difficulty is the reason that in some of the hypotheses that will be used will be found the hypothesis "no non-zero nil right ideals".

Perhaps more serious is an example due to Martindale of a prime ring in which every symmetric element is regular or nilpotent which is neither a domain nor an order in $F_2$. We shall describe this example in all detail at the end of this section.

However, all is not lost. A result due to Montgomery [11] gives a fairly definitive answer as to when this trichotomy is a consequence of the hypothesis that the elements of S (or of K) are regular or nilpotent.

We now begin the proof of Montgomery's theorem, which we described earlier as of a Faith-Utumi flavor.

Let $R$ be a prime ring with involution $*$ and let $Q = RC$ be the central closure of $R$. By Lemma 2.4.1, $*$ extends to an involution of $Q$. Suppose that $Q$ has a minimal right ideal, $eQ$, where $e^2 = e \neq 0$. Then $Q$ is primitive, acting from the right on $eQ$ with $D = eQe$ as its commuting ring, and the center of $D$ is isomorphic to $C$. We maintain this notation for a while.

LEMMA 2.5.1. Let $R, Q, e$ and $D$ be as above. If $I \neq 0$ is an ideal of $R$ such that $Ie \subset R$ then $E = eIe$ is an order in $D$, and $D = EC$.

Proof. Since $Q$ is prime, $E = eIe \neq 0$, and $EC = eICe \supset eIRCe \supset eIeQe = eIeD = D$, hence $EC = D$.

If $c_1, \ldots, c_k \in C$ then, from the nature of $C$ there exists a nonzero ideal $U \subset R$ such that $c_i U \subset R$ for $i = 1, 2, \ldots, k$. Therefore, $c_i eUIe = ec_i UIe \subset eRIe \subset eIe \subset E$, hence $U_1 = eUIe$ is an ideal of $E$ with the property that $c_i U_1 \subset E$ for all $i$, and so $c_i u = a_i \in E$ for $u \neq 0 \in U_1$. Thus $c_i = a_i u^{-1}$.

If $d \in D$, because $D = EC$, $d = \sum x_i c_i$ with $x_i \in E$, $c_i \in C$. By the above, $c_i = a_i u^{-1}$ with $a_i, u \in E$, hence $d = (\sum x_i a_i)u^{-1}$ where $\sum x_i a_i$ and $u$ are both in $E$. Thus $E$ is an order in $D$.

Let $R, Q, D, e$ be as above; then $Q$ acts faithfully and densely on a vector space $V$ over $D$. By Theorem 1.2.2, on $V$ there is a nondegenerate inner product which is Hermitian or alternate. With this notation understood we have ([9])

THEOREM 2.5.1. Let R, Q, D, e, and V be as above. Then:

1. if the form on V is Hermitian, for every integer $n \leq \dim_D V$ there exists a *-subring $R^{(n)}$ of R and an order $E^{(n)}$ in D such that:

a) $R^{(n)}$ is an order in $D_n$, and $E_n^{(n)} \subset R^{(n)} \subset D_n$.

b) the * on $R^{(n)}$ comes from an involution of transpose type on $D_n$.

2. if the form on V is alternate, D = C and for every integer $n = 2m \leq \dim_D V$ there exists a *-subring $R^{(n)}$ of R and an order $E^{(n)}$ in D such that:

a) $R^{(n)}$ is isomorphic to $E_n^{(n)}$,

b) the * on $R^{(n)}$ is the symplectic involution on $E_n^{(n)}$.

Also, if $V_0 \subset V$ is a non-degenerate subspace (i.e., the form is non-degenerate on $V_0$) of dimension n, then in both cases above $R^{(n)}$ can be chosen so that $V_0 R^{(n)} \subset V_0$ and $V_0^{\perp} R^{(n)} = 0$.

Proof. We carry out the proof first for the Hermitian case and then for the alternate one.

Case 1 The inner product on V is Hermitian.

In this case the proof of Theorem 1.2.2 revealed that Q contains a primitive, symmetric idempotent e, and Q acts faithfully and densely on V = eQ by right multiplication, and the inner product for $v, w \in V$, (v, w), is given by $(v, w) = vw^* \in D$.

Let $V_0$ be a non-degenerate subspace of V of dimension n; we can pick an orthogonal basis $\{v_1, \ldots, v_n\}$ for V over D with $d_i = (v_i, v_i) \neq 0$. If $w_i = d_i^{-1} v_i$ then $\{v_i\}$ and $\{w_j\}$ are dual bases of V, for $(v_i, w_j) = \delta_{ij}$, the Kronecker $\delta$. The elements $e_{ij}$ defined by

$ve_{ij} = (v, w_i)v_j$ are in $Q$ (see the proof of Theorem 1.2.1) and act like matrix units.

If $d \in D$ define $e_{ij} \cdot d$ by $v(e_{ij} \cdot d) = (v, w_i)dv_j$ for $v \in V$; this defines a linear transformation on $V$.

We can find an ideal $U = U^* \neq 0$ $R$ such that $0 \neq w_i^* U \subset R$, $0 \neq eU \subset R$, and $0 \neq Uv_i \subset R$ for $i = 1, 2, \ldots, n$. If $I = U^2$ then $I \neq 0$ since $R$ is prime, and $E^{(n)} = eIe$ is an order in $D$ by Lemma 2.5.1. If $x, y \in U$ let $\alpha = exye$; for $v \in V$, $v(e_{ij} \cdot \alpha) = (v, w_i)\alpha v_j = vw_i^* \alpha v_j = vw_i^* xyv_j$ since $w_i^* e = w_i$ and $ev_j = v_j$. Thus $e_{ij} \cdot \alpha$ is a right multiplication by $w_i^* xyv_j \subset w_i^* UUv_j \subset R$, hence $e_{ij} \cdot \alpha \in R$. Since $E^{(n)}$ is spanned by such $\alpha$'s, we get that $R$ contains a subring isomorphic to $E_n^{(n)}$.

Let $R^{(n)} = \{r \in R \mid r = \sum_{i,j} e_{ij} \cdot \alpha_{ij}, \ \alpha_{ij} \in D\}$. By the above, $R^{(n)} \supset E_n^{(n)}$, and since $E^{(n)}$ is an order in $D$, $R^{(n)}$ is an order in $D_n$. Since the adjoint on $D_n$ coincides with $*$ on $Q$, we get that if $r \in D^{(n)}$ then $r^* \in R^{(n)}$. This completes the Hermitian case.

<u>Case 2.</u> The inner product on $V$ is alternate.

In this case we know that $D$ is a field, so $D$ is isomorphic to $C$, and that $Q$ contains no primitive symmetric idempotent. As was shown in the proof of Theorem 1.2.2, there is a minimal right ideal $\rho = eQ$ of $Q$ such that $xx^* = 0$ for all $x \in \rho$. If $u_0 = ebe^* \neq 0$ then the inner product on $\rho$ was defined by $(v, w)u_0 = vw^*$ for $v, w \in \rho$, where $(v, w) \in D$. Note that for $\beta \in D$, from the alternate nature of the form, $\beta u_0 = u_0 \beta^*$.

If $V_0$ is a non-degenerate subspace of $V$ of dimension $n = 2m$ we can find dual bases $\{v_i\}, \{w_i\}$ of $V$ such that $(v_i, w_j) = \delta_{ij}$. As in the Hermitian case, we get matrix units in $Q$ via $ve_{ij} = (v, w_i)v_j$, and for $d \in D$, $e_{ij} \cdot d$ is defined as before.

Since $u_0 = ebe^* \neq 0$, by the minimality of $eQ$, $eQ = u_0 Q = u_0 e^* Q$. Therefore $v_i = u_0 x_i$ for $x_i \in e^* Q$ for $i = 1, \ldots, n$. We can find an ideal $U \subset R$ such that $0 \neq x_i^* U \subset R$, $0 \neq eU \subset R$, $0 \neq Ue \subset R$, and $0 \neq Uw_i \subset R$ for $i = 1, \ldots, n$. Let $I = U^2$, and $E^{(n)} = eIe$. By Lemma 2.5.1, $E^{(n)}C = D = C$ and $E^{(n)}$ is an order in $C$.

If $\alpha \in E^{(n)}$ and $v \in V$ then $v(e_{ij} \cdot \alpha) = (v, w_i)\alpha v_j = (v, w_i)\alpha u_0 x_j = (v, w_i)u_0 \alpha^* x_j = vw_i^* \alpha^* x_j$. But, by our choice of $U$, $w_i^* \alpha^* x_j \in R$ and $e_{ij} \cdot \alpha$ is merely right multiplication by this element of $R$. Hence $R$, as in case 1, contains a subring $R^{(n)}$ isomorphic to $E_n^{(n)}$. Because the involution is symplectic on $R^{(n)}$ the set of matrix units is carried into itself by $*$, hence $R^{(n)}$ is a $*$-subring of $R$. This finishes the alternate case, and so the theorem is proved.

COROLLARY. If $R$ is a prime ring with involution and satisfies a generalized polynomial identity then either $R$ satisfies a polynomial identity or, for every positive integer $n$, $R$ contains a $*$-subring $R^{(n)}$ which is a prime P.I. ring of P.I. degree at least n.

<u>Proof.</u> By Theorem 1.3.2 (Martindale's theorem), $Q = RC$ is primitive with minimal right ideal and $D$ is finite-dimensional over $C$. If $\dim_D V < \infty$ then $\dim_C V < \infty$ and then clearly $R$ satisfies a polynomial identity. Otherwise, by Theorem 2.5.1, we get $*$-subrings $R^{(n)}$ which contain orders in $D_n$, hence have P.I. degree equal to that of $D_n$, and so at least that of $C_n$, that is, $n$.

We shall now consider a special class of rings with involution. $R$ is said to be semi-normal if $xx^* = 0$ implies $x^*x = 0$ for $x \in R$. We clearly have two immediate classes of semi-normal rings:

1. $xx^* = 0$ only if $x = 0$. We described this earlier by saying that $*$ is positive definite on R.
2. $xx^* = x^*x$ for all $x \in R$. Such a ring we call normal.

We would like to prove that a certain class of semi-normal rings must either be normal or have a positive definite $*$. However, clearly the direct sum of a normal ring and one with positive definite $*$ is semi-normal. So we want to avoid direct sums; this accounts for the hypothesis of primeness in Theorem 2.5.2.

We first settle when the ring of $n \times n$ matrices over a field is semi-normal.

THEOREM 2.5.2. If $F$ is a field and $F_n$ is semi-normal, where $n > 1$, relative to an involution $*$ then either $*$ is positive definite on $F_n$ or $F_n$ is normal, $n = 2$, and $*$ is the symplectic involution.

Proof. Suppose that $*$ is not positive definite, hence $uu^* = 0$ for some $u \neq 0$. Because $ruu^*r^* = 0$ for all $r \in R$, by semi-normality of $F_n$, $u^*r^*ru = 0$.

If $*$ is of transpose type (Hermitian form) then we may suppose that $e_{ii}^* = e_{ii}$. Now since $u \neq 0$, $v = e_{ii}ue_{jj} \neq 0$ for some $i, j$ hence $v^*v = e_{jj}u^*e_{ii}e_{ii}ue_{jj} = 0$ by the above. But this reduces us to: $*$ is semi-normal on $F_2$ but not positive definite on $F_2$. We rule this out.

The $*$ induces an automorphism $-$ on $F$ and

$$\begin{pmatrix} a & b \\ c & d \end{pmatrix}^* = \begin{pmatrix} \bar{a} & \alpha^{-1}\bar{c} \\ \alpha\bar{d} & \bar{d} \end{pmatrix}$$

for some fixed $\alpha \neq 0$ in F. From

$$\begin{pmatrix} a & b \\ c & d \end{pmatrix}\begin{pmatrix} a & b \\ c & d \end{pmatrix}^* = 0$$

we get $a\bar{a} + \alpha b\bar{b} = 0$, and so

$$\begin{pmatrix} a & b \\ c & d \end{pmatrix}^*\begin{pmatrix} a & b \\ c & d \end{pmatrix} = 0,$$

which gives us $\alpha^{-1}c\bar{c} + d\bar{d} = 0$. Since $\begin{pmatrix} a & b \\ c & d \end{pmatrix} \neq 0$, not both b and c can be 0 by the above relations. Suppose $b \neq 0$; then

$$\begin{pmatrix} a & b \\ 0 & 0 \end{pmatrix}\begin{pmatrix} a & b \\ 0 & 0 \end{pmatrix}^* = 0$$

while

$$\begin{pmatrix} a & b \\ 0 & 0 \end{pmatrix}^*\begin{pmatrix} a & b \\ 0 & 0 \end{pmatrix} \neq 0 .$$

Thus we may suppose that * is symplectic. If $n > 2$ we easily get an $F_4$ such that $F_4^* \subset F_4$ and * is not positive definite on $F_4$. Let

$$A = \left(\begin{array}{cc|cc} 0 & 1 & 1 & 0 \\ 1 & 0 & 0 & 1 \\ \hline \multicolumn{2}{c|}{O} & \multicolumn{2}{c}{O} \end{array}\right); \quad \text{then} \quad A^* = \left(\begin{array}{cc|c} 0 & -1 & \\ -1 & 0 & O \\ \hline 1 & 0 & \\ 0 & 1 & O \end{array}\right) \quad \text{and } AA^* = 0$$

but $A^*A \neq 0$. We thus conclude that $n = 2$, with symplectic involution. In this case $F_2$ is trivially normal.

We now come to a pretty theorem due to Montgomery [9].

THEOREM 2.5.2. Let R be a prime, semi-normal ring. Then, either R is normal — in which case R is an order in $F_2$, F a field, with symplectic involution — or * is positive definite on R.

<u>Proof.</u> Suppose that the theorem is false. Hence there is an $x \neq 0$ in R such that $xx^* = 0$. Since $Rx \neq 0$, we can in fact choose x in $R^2$. If $R^2$ is an order in $F_2$ then, trivially, R is an order in $F_2$. Since

we are denying the theorem, by Theorem 2.2.1, $R^2$ must contain a symmetric element $a \neq 0$ such that $a^2 = 0$.

If $r \in R$ then $0 = raar^* = (ra)(ra)^*$, hence, by semi-normality, $ar^*ra = (ra)^*(ra) = 0$; linearizing on $r$ we obtain $a(xy + y^*x^*)a = 0$ for all $x, y \in R$. Thus $a(u + u^*)a = 0$ for all $u \in R^2$. Letting $u = xay$ where $x, y \in R^2$, we obtain, as usual, that $axaya = -ayaxa$ for all $x, y \in R$.

If char $R \neq 2$, this last relation gives $2(ax)^3 = 0$ and so $(ax)^3 = 0$; by Lemma 2.1.1 this gives $R^2$ a nilpotent ideal, contrary to the primeness of $R$. Hence we may suppose that char $R = 2$. In that case, $axaya = ayaxa$ for all $x, y \in R^2$. By Theorem 1.3.1, $Q = R^2C$ is primitive with minimal right ideal $V$ and commuting ring $C$; the proof of Theorem 1.3.1 also shows that $a$ is of finite rank (in fact, of rank 1). Thus $Va$ is finite dimensional over $C$.

Let $V_0$ be any non-degenerate finite dimensional subspace of $V$ such that $Va \subset V_0$ and $V_0^{\perp} a = 0$. Because $a^2 = 0$ and $V_0$ is non-degenerate we must have $n = \dim_C V_0 \geq 2$. We view $a$ as a matrix in $C_n$ acting on $V_0$.

By Theorem 2.5.1, for some order $E$ in $C$, $E_n \subset R^2$ and $E_n$ acts on $V_0$. Hence, if $x \in C_n$, $x\alpha \in E_n \subset R^2$ for some $\alpha \neq 0$ in $E$. If $xx^* = 0$ then $(x\alpha)(x\alpha)^* = \alpha\bar{\alpha}xx^* = 0$; since $x\alpha \in R^2$, we have $(x\alpha)^*(x\alpha) = 0$, and so $x^*x = 0$ results. Thuc $C_n$ is semi-normal; since $a \in C_n$ and $0 = a^2 = aa^*$, $*$ is not positive definite on $C_n$. By Lemma 2.5.2, $n = 2$ (and $*$ is symplectic on $C_2$). Therefore the only non-degenerate finite dimensional subspaces containing $Va$ are

2-dimensional over C. This forces $\dim_C V = 2$; hence $R^2$, and so R, is an order in $C_2$. With this the theorem is proved.

We can not expect a good subdirect product of a semi-prime ring which is semi-normal into normal and positive definite constituents. To see this, let D be a non-commutative division ring and $D^0$ is opposite ring. If $R = D \oplus D^0$ then, relative to the exchange involution, R is *-subdirectly irreducible. So R has no subdirect product decomposition into *-images. Yet R is clearly semi-normal.

We now apply the two theorems proved above to the kind of theorems we have considered earlier, namely those in which we impose some decency condition on S, T, K, or $K_0$.

THEOREM 2.5.3. Let R be a prime ring with involution whose central closure has a minimal right ideal. If every trace (or, if every skew trace) of R is regular or nilpotent then either R is a domain or is an order in $F_2$, F a field.

Proof. By Theorem 2.5.1, R contains a subring $E_n$, for every $n \leq \dim_D V$, where E is an order in D.

If the inner product on V is alternate then D = C and * is symplectic on $E_n$. If $n > 2$, let

$$x = \left(\begin{array}{cc|c} \alpha & 0 & O \\ 0 & 0 & \\ \hline O & & O \end{array}\right)$$

where $\alpha \neq 0 \in E$; then $a = x \pm x^*$ is a trace or skew trace in R, respectively, but since

$$a = \left(\begin{array}{cc|c} \alpha & 0 & O \\ 0 & \pm\alpha & \\ \hline O & & O \end{array}\right) ,$$

a is neither nilpotent nor regular. Thus $n = 2$ follows and so $\dim_C V = 2$ and $R$ is an order in $C_2$.

If the form on $V$ is Hermitian and $n > 2$, if $\alpha \neq 0 \in E$ then for

$$x = \left(\begin{array}{cc|c} 0 & \alpha & \\ 0 & 0 & O \\ \hline O & & O \end{array}\right), \quad x^* = \left(\begin{array}{cc|c} 0 & 0 & \\ \gamma\bar{\alpha}\delta & 0 & O \\ \hline O & & O \end{array}\right)$$

for some $\gamma, \delta$ nonzero in $D$, not depending on $\alpha$. But $x^*$ is in $R$ and so

$$x \pm x^* = \left(\begin{array}{cc|c} 0 & \alpha & \\ \gamma\bar{\alpha}\delta & 0 & O \\ \hline O & & O \end{array}\right)$$

is neither regular nor nilpotent; thus $\dim_D V = 2$ and $Q = RC = D_2$.

We claim that $D$ is a field; if not, $E$ as an order in $D$ cannot be commutative. There exists $\beta \in D$ such that for all $\alpha \in E$, if $x = \begin{pmatrix} \alpha & 0 \\ 0 & 0 \end{pmatrix}$

then $x^* = \begin{pmatrix} \beta\bar{\alpha}\beta^{-1} & 0 \\ 0 & 0 \end{pmatrix}$; we know that $x^* \in R$ since $x \in E_2 \subset R$.

Since $x \pm x^*$ is nilpotent or regular we get that $\alpha \pm \beta\bar{\alpha}\beta^{-1} = 0$ for all $\alpha \in D$. If we were talking about skew traces, we would have $\alpha\beta = \beta\bar{\alpha}$ for all $\alpha \in D$ and so $\bar{\alpha} = \beta^{-1}\alpha\beta$ is an automorphism of $D$. But $\bar{\ }$ is an anti-automorphism of $D$. $D$ therefore must be commutative.

If we were talking about traces, then $\alpha + \beta\bar{\alpha}\beta^{-1} = 0$. Using $\alpha = 1$ we get $2 = 0$ and so $\operatorname{char} D = 2$. This puts us back to the case of skew traces already settled, so $D$ is a field.

As an immediate corollary we have the prime case of Theorem 2.4.1.

COROLLARY 1. If $R$ is prime and every non-zero skew trace (or, if every non-zero trace) in $R$ is regular, then $R$ is a domain or is an order in $F_2$, $F$ a field.

Proof. We give the proof for skew traces; that for traces is very similar.

By Theorem 2.2.1, the result is correct unless there is an element $a = a^* \neq 0$ in $R$ such that $a^2 = 0$. Since $aK_0a \subset K_0$ consists of nilpotent elements, by our hypothesis, $aK_0a = 0$. Thus $axa = ax^*a$ for all $x \in R$. Replacing $x$ by $xay$ in this relation — $axa = ax^*a$ — yields $axaya = ayaxa$ for all $x, y \in R$. By Theorem 1.3.1, $RC$ is primitive with minimal right ideal. Applying Theorem 2.5.3 we have the corollary.

The second consequence of the theorem is a result of Montgomery [11].

COROLLARY 2. Let $R$ be prime in which every trace (or, every skew trace) is nilpotent or regular. Suppose that $R$ is not an order in $F_2$; then $R$ is a domain if and only if whenever $xx^* = 0$ in $R$ then $x^*x = 0$ (i.e., $R$ is semi-normal).

Proof. If $R$ is a domain $xx^* = 0$ only if $x = 0$, and so $x^*x = 0$. On the other hand, if $R$ is semi-normal, by Theorem 2.5.2 $*$ is positive definite on $R$. Thus no symmetric or skew element in $R$ can b be nilpotent. Applying Corollary 1 we have Corollary 2.

For several theorems of the flavor of Corollary 2, once we had the result for the prime case we were able to go on to more general cases of the theorem. This is true here also. The manner in which we do it is reminiscent of the passage made for these other results earlier. The result we prove is due to Montgomery [11].

THEOREM 2.5.3. Let R be a non-commutative ring having no non-zero nil right ideals. Suppose that either

1. every element in T is regular or nilpotent, or
2. every element in $K_0$ is regular or nilpotent.

Then R is a domain, a subdirect sum of a domain and its opposite, or an order in $F_2$ if and only if whenever $xx^* = 0$ in R then $x^*x = 0$.

<u>Proof.</u> The necessity of the condition is clear. Now to the sufficiency. If R is prime, then the result is correct by Corollary 2 to Theorem 2.5.3. Suppose then that R is not prime; hence we have two ideals $A \neq 0$, $B \neq 0$ of R such that $AB = 0$.

As we have always done in this procedure, our first step is to show that $A \cap A^* = 0$. Now if $A \cap A^* = U \neq 0$ then since $U^* = U$ and $UB = 0$, we have $B^*U = 0$. However, R is semi-prime, hence $UB^* = 0$. Thus $U(B + B^*) = 0$. Since $U \neq 0$, $B + B^*$ can have no regular elements; thus every $t \in T \cap (B+B^*)$ or every $k \in K_0 \cap (B+B^*)$ must be nilpotent, according to which hypothesis — on T or $K_0$ — that we are using. In either case, by Lemma 2.3.1, $B+B^*$ would have a non-zero nil right ideal — which would give us a non-zero nil right ideal in R, which is not possible — or $B+B^*$ would have to be commutative. In this latter case $B+B^*$ would have to be in the center of R, and so would annihilate the commutator ideal $\mathfrak{C}$ of R. Since $\mathfrak{C}^* = \mathfrak{C}$ and $\mathfrak{C}(B+B^*) = 0$ we could repeat the argument used on $B+B^*$ to get that $\mathfrak{C}$ is in the center of R, and so $\mathfrak{C}$ would annihilate $\mathfrak{C}$, giving us the contradiction $\mathfrak{C}^2 = 0$

and $\zeta \neq 0$ in the semi-prime ring R. We have thus shown that $D = A \cap A^*$ must indeed be 0, and so $AA^* = A^*A = 0$.

We claim that A is a domain. For if $ab = 0$ where $a \in A$, $0 \neq b \in A$ then $(a \pm a^*)b = 0$, hence neither $a + a^*$ nor $a - a^*$ can be regular. According as our hypothesis is on T or $K_0$, we have $a + a^*$ or $a - a^*$ nilpotent; either gives, since $aa^* = a^*a = 0$, that a is nilpotent. If $r \in R$ then $ra \in A$ and $(ra)b = 0$; thus ra is nilpotent. Since Ra is a nil left ideal of R, by our hypothesis on R we get that $a = 0$. Thus A is a domain.

We further claim that R has no nilpotent elements. For, if $x^2 = 0$ then $xAx \subset A$ consists of nilpotent elements (since $(xAx)^2 = 0$). But A is a domain, hence $xAx = 0$ and so $xA = 0$. Similarly, $xA^* = 0$; hence $x(A + A^*) = 0$. But no element $a \pm a^* \neq 0$ in A can be nilpotent, so one of these is regular (according to our hypothesis). Thus $x = 0$.

Since R has no nilpotent elements, our hypothesis reduces to: every non-zero element in T or $K_0$ is regular. Applying Theorem 2.2.2 or Theorem 2.4.1 gives us the result.

Note that the proof gave a little more than we claimed, namely: if R is a non-commutative ring with no nil right ideals in which every element in T is regular or nilpotent, or every element in $K_0$ is regular or nilpotent then, if R is <u>not</u> prime it must be a subdirect sum of a domain and its opposite. No use was needed in this case of $xx^* = 0$ implies $x^*x = 0$ — that follows free of charge.

If R is a semi-prime Goldie ring, for instance, then every left zero-divisor in R must be a right zero-divisor. Also, by a result of

Lanski [7], R has no non-zero nil right ideals. Thus, here, if we impose the condition of regularity or nilpotence on T or $K_0$ we get the usual trichotomy for R.

We close this section and chapter with a discussion of the example of Martindale which we mentioned earlier. The argument we give is due to Estes and Lanski (unpublished). Another approach to the verification that the example does what it does can be found in [2].

Let F be a field and F[x, y] the free algebra in x and y over F. F[x, y] has a natural involution: the * of a word in x and y is the word read backward; this extends by linearity to F[x, y]. The ideal $(x^2)$ is a *-ideal of F[x, y], hence $R = \frac{F[x,y]}{(x^2)}$ is a ring with involution. It is a prime ring, and as we shall see, in it every symmetric element is regular or has square 0. Yet R is neither a domain nor an order in $F_2$.

If $A = F[x,y]$ and $r \in A$, let $\delta(r)$ be defined by: $\delta(\alpha) = 0$ for $\alpha \in F$, $\delta(r)$ = maximum of the degrees of the monomials appearing in a minimal representation of r as a linear combination of monomials.

The following four statements are obvious:

(1) $xr \in (x^2)$ implies $r \in xA + (x^2)$.

(2) $yr \in xA + (x^2)$ implies $r \in (x^2)$.

(3) $rx \in xA + (x^2)$ implies $r \in xA + Ax + (x^2)$.

(4) $rx \in (x^2)$ implies $r \in Ax + (x^2)$.

We shall refer to these statements by their numbers in the proofs that are to come.

We first prove the

PROPOSITION 1. If $ab \in xA + (x^2)$ then $a \in xA + (x^2)$ or $b \in xA + (x^2)$.

Proof. Suppose that the proposition is false; pick $a \in A$ with $\delta(a)$ minimal for which the proposition fails. For this $a$ let $b \in A$ be chosen with $ab \in xA + (x^2)$, with $\delta(b)$ minimal, for which $b \notin xA + (x^2)$.

We write $a = \alpha + xa_1 + ya_2$ and $b = \beta + xb_1 + yb_2$ where $\alpha, \beta \in F$. Since $ab \in xA + (x^2)$ we have $\alpha\beta = 0$, so $\alpha = 0$ or $\beta = 0$.

If $\alpha = 0$ we get $ya_2b \in xA + (x^2)$ so, by (2), $a_2b \in (x^2)$. So, since $\delta(a_2) < \delta(a)$, either $b \in xA + (x^2)$ — which we supposed it isn't — or $a_2 \in xA + (x^2)$. In this second case, write $a_2 = xa_3 + t$ where $t \in (x^2)$ and so $xa_3b \in (x^2)$ since $a_2b \in (x^2)$. By (1) we have $a_3b \in xA + (x^2)$; since $b \notin xA + (x^2)$ and $\delta(a_3) < \delta(a_2) < \delta(a)$ we must have $a_3 \in xA + (x^2)$. Hence $xa_3 \in (x^2)$ and so $a_2 \in (x^2) \subset xA + (x^2)$, a contradiction. Thus the proposition is correct if $\alpha = 0$.

Suppose, then, that $\beta = 0$; thus $b = xb_1 + yb_2$. Since $ab \in xA + (x^2)$, we have $(\alpha + ya_2)b \in xA + (x^2)$. If $\delta(\alpha + ya_2) < \delta(a)$ then we are done. So we have that $\delta(\alpha + ya_2) = \delta(a)$. Multiply out $(\alpha + ya_2)b \in xA + (x^2)$; we get $y(\alpha b_2 + a_2xb_1 + a_2yb_2) \in xA + (x^2)$. By (2), $\alpha b_2 + a_2xb_1 + a_2yb_2 \in (x^2)$, and since $\alpha \neq 0$, we have $b_2 = a_2u + v$ where $v \in (x^2)$. Thus, since $(\alpha b_2 + a_2xb_1 + a_2yb_2) \in (x^2)$, using the value $b_2 = a_2u + v$ gives us that $a_2(\alpha u + xb_1 + yb_2) \in (x^2)$. Since $\delta(a_2) < \delta(a)$, either $a_2 \in xA + (x^2)$ or $\alpha u + xb_1 + yb_2 \in xA + (x^2)$. If $a_2 \in xA + (x^2)$ then $a_2 = xa_4 + t$ where $t \in (x^2)$, and $\alpha u + xb_1 + yb_2 \notin xA + (x^2)$ then $xa_4(\alpha u + xb_1 + yb_2) \in (x^2)$, so by (1), $a_4(\alpha u + xb_1 + yb_2) \in xA + (x^2)$. Since $\delta(a_4) < \delta(a_2) < \delta(a)$, and since $\alpha u + xb_1 + yb_2 \notin xA + (x^2)$, we must have $a_4 \in xA + (x^2)$ and so $xa_4 \in (x^2)$, giving us that $a_2 \in (x^2)$. Thus, since $a = \alpha + xa_1 + ya_2$ and $ab \in xA + (x^2)$, we have $\alpha b \in xA + (x^2)$ with $\alpha \neq 0 \in F$. The net result is $b \in xA + (x^2)$, a contradiction.

In consequence we must assume, above, that $\alpha u + xb_1 + yb_2 \in xA + (x^2)$, and so $\alpha u + yb_2 \in xA + (x^2)$. However, $b_2 = a_2u + v$ where $v \in (x^2)$; we thus have $(\alpha + ya_2)u \in xA + (x^2)$. Now $\delta(\alpha + ya_2) = \delta(a)$ but $\delta(u) < \delta(b_2) < \delta(b)$. By our choice of a and b we must have either $\alpha + ya_2 \in xA + (x^2)$ — which gives $a = \alpha + ya_2 + xa_1 \in xA + (x^2)$, a desired outcome — or $u \in xA + (x^2)$. Thus $u \in xA + (x^2)$.

We write $u = xu_1 + w$ where $w \in (x^2)$. Since $(\alpha + ya_2)u \in xA + (x^2)$ we get that $ya_2xu_1 \in xA + (x^2)$. By (2) we conclude that $a_2xu_1 \in (x^2)$. Since $\delta(a_2x) \leq \delta(a)$ and $\delta(u_1) < \delta(b)$ we have $u_1 \in xA + (x^2)$ or $a_2x \in xA + (x^2)$. If $u_1 \in xA + (x^2)$ then $u = xu_1 + w \in (x^2)$ and so $b_2 = a_2u + v \in (x^2)$; this gives that $b = xb_1 + yb_2$ must be in $xA + (x^2)$, a contradiction. Consequently, $a_2x \in xA + (x^2)$. By (3) we must have that $a_2 = c_1x + xc_2 + d$ where $d \in (x^2)$. Hence $a_2xu_1 \in (x^2)$ implies that $xc_2xu_1 \in (x^2)$, whence, by (1), $c_2xu_1 \in xA + (x^2)$. Since $\delta(c_2x) \leq \delta(a_2) \leq \delta(a)$ and $\delta(u_1) < \delta(b)$, and since $u_1 \notin xA + (x^2)$ — as we saw above — we must have $c_2x \in xA + (x^2)$, hence by (3) $c_2 \in xA + Ax + (x^2)$. This gives us that $xc_2 \in Ax + (x^2)$ and so $a_2 = c_1x + xc_2 + d \in Ax + (x^2)$. We write $a_2 = ex + f$ where $f \in (x^2)$; thus $b = xb_1 + yb_2 = xb_1 + y(a_2u + v) = xb_1 + y((ex + f)(xu_1 + w)) \in xA + (x^2)$ since $(ex + f)(xu_1 + w) \in (x^2)$ because $w, f \in (x^2)$. This finishes the proof of the proposition.

With this proposition proved we can prove that the zero divisors in $R = \frac{A}{(x^2)}$ are of the desired form, and that every symmetric element in R is regular or has square 0. We prove the result which immediately implies this statement.

PROPOSITION 2. If $a, b \in A$ are such that $a \notin (x^2)$, $b \notin (x^2)$ but $ab \in (x^2)$ then $a \in Ax + (x^2)$ and $b \in xA + (x^2)$.

Proof. Since $ab \in (x^2) \subset xA + (x^2)$, by Proposition 1, $a \in xA + (x^2)$ or $b \in xA + (x^2)$.

If $b \in xA + (x^2)$ then $b = xb_1 + t$, $t \in (x^2)$; since $b \notin (x^2)$, $b_1 \notin xA + (x^2)$. Since $ab \in (x^2)$ we have $axb_1 \in (x^2) \subset xA + (x^2)$, and since $b_1 \notin xA + (x^2)$, $ax \in xA + (x^2)$. By (3), $a \in xA + Ax + (x^2)$, so $a = xa_1 + a_2x + s$ where $s \in (x^2)$. Since $axb_1 \in (x^2)$ we have $xa_1xb_1 \in (x^2)$, so by (1), $a_1xb_1 \in xA + (x^2)$, and since $b_1 \notin xA + (x^2)$, by Proposition 1, $a_1x \in xA + (x^2)$. Thus by (3), $a_1 \in xA + Ax + (x^2)$, hence $xa_1 \in Ax + (x^2)$; since $a = xa_1 + a_2x + s$, we get $a \in Ax + (x^2)$ as required.

On the other hand, if $a \in xA + (x^2)$ write $a = xa_1 + t$, $t \in (x^2)$. Since $a \notin (x^2)$, $a_1 \notin xA + (x^2)$. Since $ab \in (x^2)$ we have $xa_1b \in (x^2)$ and so, by (1), $a_1b \in xA + (x^2)$. Since $a_1 \notin xA + (x^2)$, by Proposition 1, $b \in xA + (x^2)$. By the above we have that $a \in Ax + (x^2)$, and the proposition is proved.

Bibliography

1. W. Baxter. Lie simplicity of a special class of associative rings II. Trans. AMS 87 (1958): 63-75.

2. P. M. Cohn. Prime rings with involution all of whose symmetric elements are nilpotent or regular. Proc. AMS 40 (1973): 91-92.

3. I. N. Herstein. Topics in Ring Theory. Univ. of Chicago Press, 1969.

4. I. N. Herstein. On rings with involution. Canadian Math. Jour. 26 (1974): 794-799.

5. I. N. Herstein and S. Montgomery. Invertible and regular elements in rings with involution. Jour. Algebra 25 (1973): 390-400.

6. C. Lanski. Rings with involution whose symmetric elements are regular. Proc. AMS 33 (1972): 264-270.

7. C. Lanski. Nil subrings of Goldie rings are nilpotent. Canadian Math. Jour. 21 (1969): 904-907.

8. K. McCrimmon. On Herstein's theorems relating Jordan and associative algebras. Jour. Algebra 13(1969): 382-392.

9. S. Montgomery. A structure theorem and a positive definiteness condition in rings with involution, (to appear).

10. S. Montgomery. Rings of quotients for a class of special Jordan rings. Jour. Algebra 31(1974): 154-165.

11. S. Montgomery. Rings with involution in which every trace is nilpotent or regular. Canadian Math. Jour. 26 (1974): 130-137.

12. J. M. Osborn. Jordan algebras of capacity two. Proc. Nat. Acad. Sci. USA 57 (1967): 582-588.

3

# COMMUTATIVITY THEOREMS

In the early stages of general ring theory, some of the striking successes of that theory were theorems which asserted that when the elements of a ring were subjected to certain types of algebraic conditions, the ring itself had to be commutative or almost commutative. A good cross-section of such results, and the techniques needed to obtain them, can be found in [9] and [3].

It is reasonable to try to transfer these kinds of theorems to the setting of rings with involution. Firstly, these types of questions arise naturally in the theory of Jordan algebras (see, for instance [16]). Secondly, from the point of view of technique and approach, the problems that present themselves, in this way, for rings with involution offer interesting challenges. The reason is simply that the usual ring theoretic methods devised earlier are not adequate to handle these new situations.

Generally speaking, the set-up studied will be as follows: suppose that we impose the kind of "commutativity hypotheses" mentioned above on the set of symmetric, or on the set of skew, elements or on appropriate subsets of these; what global effect do these have on the rings themselves?

As we shall see in this chapter, there are suitable analogs of these general commutativity theorems for rings with involution. However, the general conclusion will not be that the ring in question is commutative or almost commutative. This is too much to expect, for there exist too many decent counter-examples to such hopes. So, to call such results "commutativity theorems" is probably a misnomer. Be that as it may, we shall show that the imposition of such hypotheses on rings with involutions renders these rings to be very special, indeed, in structure.

## 1. Division Rings

A famous result of Jacobson says that if R is a ring in which $x^{n(x)} = x$, $n(x) > 1$, for every $x \in R$, then R must be commutative.

In general, if we insist, say, that $s^{n(s)} = s$, $n(s) > 1$, for all $s \in S$ we cannot conclude that R is commutative. In fact we cannot even conclude that any two elements of S must commute. For instance, if F is the field of integers modulo 3, and $R = F_2$ relative to transpose as the involution, we can readily verify that $s^{n(s)} = s$ for all $s \in S$, with $n(s) > 1$. Yet, here, symmetric elements need not commute. There is no doubt that this ring R is very decent — it is, after all, a finite, simple ring. So we must expect answers of a different kind.

The story for skew elements will turn out to be more decisive in this regard, and we shall see that in a large class of situations we will be able to say that any two skew elements must commute.

For division rings with involution the theorem of Jacobson goes over in its entirety. It is this sort of results — which come from a paper by Herstein and Montgomery [8] — that we discuss in this section. We begin with the case of symmetric elements.

Let D be a division ring with involution * and suppose that for every $s \in S$ there is an integer $n(s) > 1$ such that $s^{n(s)} = s$. We propose to show that D is commutative.

First note that D must have finite characteristic. For, if $s \neq 0$ is in S and $s^n = s$, $(2s)^m = 2s$, where $n > 1$, $m > 1$, then $s^q = s$ and $(2s)^q = 2s$ where $q = (m-1)(n-1) + 1 > 1$. These give us that $(2^q - 2)s = 0$, hence $\operatorname{char} D = p \neq 0$.

Let P be the prime field of p elements and Z, as usual, the center of D; thus $P \subset Z$.

We begin with

LEMMA 3.1.1. If $x \in D$ is such that $xx^* = x^*x$ then x is algebraic over P, and $x^{n(x)} = x$ where $n(x) > 1$.

Proof. Since $xx^* = x^*x$ we immediately have that $x + x^*$ and $x^*x$ both commute with x and with each other. Since $x + x^*$ and $x^*x$ are in S, by our assumption on D they must be algebraic over P. Thus $F = P(x + x^*, x^*x)$ is a finite field. Since x centralizes F and $x^2 - \alpha x + \beta = 0$ where $\alpha = x + x^*$, $\beta = x^*x$ are in F, we have that x is algebraic over F. Thus $F(x)$ is a finite field, and so x is algebraic over P. Because $P(x)$ is a finite field, $x^{n(x)} = x$ for some $n(x) > 1$.

If $z \in Z$ then certainly $zz^* = z^*z$, hence we have the

COROLLARY. Z is algebraic over P.

A basic step in our discussion is

LEMMA 3.1.2. If char $D \neq 2$ and if $a^2 \in Z$ for some $a \in S$, then $a \in Z$.

<u>Proof</u>. We may assume that $a \neq 0$. If $b \in S$ then $c = ab - ba$ satisfies:

1. $c^* = -c$

2. $ac + ca = 0$.

Since $c^2 \in S$, $c^2$ is algebraic over $P$, hence $c$ is algebraic over $P$. So $c^n = c$, say, with $n > 1$. Also, since $a \in S$, $a^m = a$ with $m > 1$. From these and the fact that $ac = -ca$ we trivially get that $a$ and $c$ generate a finite ring over $P$, hence a finite division ring, $D_0$. By Wedderburn's theorem on finite division rings, $D_0$ is commutative. But $a, c$ are both in $D_0$; hence $ac = ca$. However, we know that $ac = -ca$. The net outcome of this is that $2ac = 0$, and since char $D \neq 2$, and $a \neq 0$, we must have $c = 0$. In short, $a$ must commute with all symmetric elements in $D$.

But, by Theorem 2.1.6, $\overline{S} = D$ unless $\dim_Z D \leq 4$. If $\overline{S} = D$ then, since $a$ centralizes $S$, $a$ must be in $Z$. If $\dim_Z D \leq 4$, then, since $Z$ is algebraic over $P$ by the corollary to Lemma 3.1.1, $D$ must be algebraic over $P$. By Jacobson's theorem, $D$ must be commutative, in which case $a$ certainly is in $Z$. The lemma is now proved.

We are now in a position to prove the first of our commutativity theorems.

THEOREM 3.1.1. Let $D$ be a division ring with involution in which $s^{n(s)} = s$, $n(s) > 1$, for every $s \in S$. Then $D$ is commutative. In fact, $D$ is algebraic over the prime field with $p$ elements, where char $D = p \neq 0$.

Proof. In order to show that $D$ is commutative it is enough for us to show that $D$ is algebraic over $P = GF(p)$, for the commutativity would then follow by invoking Jacobson's theorem.

If $S \subset Z$ then $x + x^* \in Z$ and so commutes with $x$; we would then have that each $x \in D$ is algebraic over $P$ and we would be done. So we may assume that for some $a^* = a$ in $D$, $a \notin Z$. Since $a$ is algebraic over $P$ — $a^{n(a)} = a$ — and $a \notin Z$ there is an element $b \in D$ such that $bab^{-1} = a^i \neq a$ (see [3] for this special case of the Skolem-Noether theorem). The element $b$ cannot have finite multiplicative order, otherwise $a$ and $b$ would generate a finite ring over $P$ — because of the relations $a^{n(a)} = a$, $b^{n(b)} = b$, $bab^{-1} = a^i$ — which, by Wedderburn's theorem would be commutative. This would contradict $ab \neq ba = a^i b$.

Apply $*$ to the relation $bab^{-1} = a^i$; since $a^* = a$ we get $(b^*)^{-1}ab^* = a^i = bab^{-1}$. Hence $\lambda = b^*b$ commutes with $a$. Since $\lambda^* = \lambda$, $\lambda^k = \lambda$ for some $k > 1$. We claim that $k$ is odd. Otherwise, if $\mu = \lambda^{k/2}$, then $\mu^* = \mu$, $\lambda = \mu^2 = \mu^*\mu$, $\mu$ commutes with $a$, and $b^*b = \mu^*\mu$. But then, if $c = b\mu^{-1}$, we have $c^*c = 1 = cc^*$, hence, by Lemma 3.1.1, $c$ is algebraic over $P$, and so $c^m = c$ for some $m > 1$. But $cac^{-1} = b\mu^{-1}a\mu b^{-1} = bab^{-1} = a^i$, since $\mu a = a\mu$. Since $c$ has finite multiplicative order, and $cac^{-1} = a^i \neq a$, as we saw earlier, this leads to a contradiction. We must therefore have that $k$ is odd. In particular, D cannot have characteristic 2, for in the finite field $P(\lambda)$, if char D = 2, $\lambda^{2^n} = \lambda$ would hold. The order of $\lambda$ is, of course, then even.

Suppose that $\lambda$ has order $2^u v$ where v is odd. We can write $\lambda$ as $\lambda = \lambda_1\lambda_2$ where $\lambda_1, \lambda_2$ are powers of $\lambda$ and $\lambda_1^{2^u} = 1$, $\lambda_2^v = 1$; since $\lambda_1, \lambda_2$ are powers of $\lambda$, $\lambda_1^* = \lambda_1$, $\lambda_2^* = \lambda_2$.

But, $\lambda_2^{v+1} = \lambda_2$, and since $v+1$ is even, $\lambda_2 = \mu^2 = \mu^*\mu$ where $\mu = \lambda_2^{(v+1)/2}$. Let $c = b\mu^{-1}$; then $c^*c = \mu^{-1}b^*b\mu^{-1} = \mu^{-1}(\lambda_1\lambda_2)\mu^{-1} = \mu^{-2}\lambda_1\lambda_2 = \lambda_1$. Hence $(c^*c)^{2^u} = 1 \in Z$; since char D $\neq$ 2, invoking Lemma 3.1.2 repeatedly, we get that $c^*c \in Z$, and so $c^*c = cc^*$. By Lemma 3.1.1 this forces c to be algebraic over P and of finite multiplicative order. However, $cac^{-1} = b\mu^{-1}a\mu b^{-1} = bab^{-1} = a^i \neq a$ since $\mu$ commutes with a. We saw that this is not possible earlier. Hence the theorem is proved.

We now prove the skew analog of Theorem 3.1.1. However, we cannot conclude from $a^{n(a)} = a$, for all $a \in K$, that D is algebraic over a finite field. For, if D is any field, and $a^* = a$ for all $a \in D$ then, vacuously, $a^{n(a)} = a$ for all skew elements. However, we shall show that D must be commutative. This is

THEOREM 3.1.2. Let D be a division ring with involution in which $a^{n(a)} = a$, $n(a) > 1$, for every $a \in K_0$. Then D must be commutative. If $K_0 \neq 0$ then D must be algebraic over the prime field P of p elements.

Proof. If $x^* = x$ for all $x \in D$ then D is a field, and nothing more can be said.

Suppose, then, that $K_0 \neq 0$. If $k \neq 0 \in K_0$, then $2k \in K_0$ and from $k^n = k$, $(2k)^m = 2k$, with $n > 1$, $m > 1$, we conclude that char $D = p \neq 0$.

If $K_0 \cap Z \neq 0$, let $\mu \neq 0 \in K_0 \cap Z$ and $0 \neq s \in S$. Then $\mu s^2 = s\mu s$ is in $K_0$, hence $(\mu s^2)^n = \mu s^2$ for some $n > 1$. Since $\mu \in K_0$, $\mu^m = \mu$ for some $m > 1$. Thus, if $q = (m-1)(n-1) + 1$ we get, from $(\mu s^2)^q = \mu s^2$ and $\mu^q = \mu$, that $\mu(s^{2q} - s^2) = 0$, hence $s^{2q} = s^2$, and so, $s^{2q-1} = s$. Invoking Theorem 3.1.1 we would have that $D$ is a field algebraic over $P = GF(p)$.

Let $k \neq 0$ be in $K_0$. Then $C_D(k) = \{x \in D \mid xk = kx\}$ is a subdivision ring of $D$ invariant re *. Moreover $k \in Z(C_D(k)) \cap K_0$, hence by the above, $C_D(k)$ is a field algebraic over $P$. Since $Z \subset C_D(k)$, $Z$ is algebraic over $P$.

Because $k^{n(k)} = k$, $n(k) > 1$, $k$ is algebraic over $P$, hence over $Z$. But $C_D(k)$ is a maximal subfield of $D$. Consequently, since in this case $C_D(k) = Z(k)$ by the double centralizer theorem, $D$ must be finite dimensional over $Z$. Therefore $D$ is algebraic over $P$. By Jacobson's theorem $D$ must be a field.

A very special case of the theorem is

COROLLARY 1. If $D$ is a division ring with involution in which $a^{n(a)} = a$, $n(a) > 1$, for all $a \in K$ then $D$ is a field. If * is not the identity map on $D$ then $D$ is algebraic over $P = GF(p)$ for some prime $p$.

COROLLARY 2. If $D$ is a division ring with involution in which $t^{n(t)} = t$ for all $t \in T$, then $D$ is a field. If $T \neq 0$ (which is automatic, unless char $D = 2$ and $*$ is the identity map on D) then D is algebraic over $P = GF(p)$ for some prime p.

Proof. If char $D \neq 2$, then $T = S$ and this corollary is merely Theorem 3.1.1. If char $D = 2$, the corollary is the special case of Theorem 3.1.2 in characteristic 2.

## 2. More on Division Rings

We now examine the influence of other commutativity hypotheses on division rings. A theorem of Kaplansky [11] states that if D is a division ring in which $x^{n(x)} \in Z$, $n(x) > 1$, for all $x \in D$ then D is commutative. We begin this section by looking at division rings with involution where powers of the symmetric elements, or powers of the skew elements, are central.

To get started on these problems we need a variant of the Jacobson-Noether theorem which assures us that in algebraic division algebras there are lots of separable elements. The result is due to Chacron [1].

LEMMA 3.2.1. Let D be a division ring with involution, of characteristic not 2, in which every element in S is algebraic over Z. Then, either $S \subset Z$ (and so, $\dim_Z D \leq 4$) or there is an element $a \in S$, $a \notin Z$, which is separable over Z.

Proof. Suppose that $S \not\subset Z$, and suppose that the lemma is false. Then every $s \in S$, $s \notin Z$, is purely inseparable over Z. Hence there is an $a \in S$, $a \notin Z$ such that $a^p \in Z$, where $p = \text{char } D \neq 0$.

Since $S \not\subset Z$, $S$ generates $D$; thus, because $a \notin Z$, there exists an element $s \in S$ such that $sa - as \neq 0$. But

$$\underbrace{[\cdots[s,a],a]\cdots]a]}_{p\text{-times}} = sa^p - a^p s = 0 .$$

Thus there is some $i > 1$ such that $x = [\cdots\overbrace{[s,a]a]\cdots a]}^{i-1 \text{ times}}] \neq 0$ but $xa - ax = 0$. From its form, $x = ya - ay$ where $y^* = y$ or $y^* = -y$ (depending on the parity of $i$). At any rate, since $x$ commutes with $a$ we can write $x = ta$ where $ta = at$. Thus $ta = x = ya - ay$, which gives us $a = ua - au$ where $u = t^{-1}y$. Applying $*$ to this we have $a = au^* - u^*a$ and so $2a = (u - u^*)a - a(u - u^*)$, whence $a = ka - ak$ where $-k^* = k = \frac{u - u^*}{2}$. Now, $k^2 \in S$, hence $k^{2p^n} \in Z$ for some $n \geq 1$. However, from $a = ka - ak$, $a^{-1}ka = 1 + k$ and so $(1+k)^{2p^n} = a^{-1}k^{2p^n}a = k^{2p^n}$. Expanding this we have $1 + 2k^{p^n} + k^{2p^n} = k^{2p^n}$, and so $2k^{p^n} = -1$. This last relation is impossible since, because $p$ is odd, $k^{p^n}$ is skew whereas $-1/2$ is symmetric. The lemma is proved.

In characteristic 2 the result is false. We can easily construct a 4-dimensional division algebra with involution, of characteristic 2, where $S \not\subset Z$ but $s^2 \in Z$ for all $s \in S$. As we shall soon see, this is the only possible counter-example.

LEMMA 3.2.2. Let char $D = 2$ and suppose that there is a $\lambda \in Z$ with $\lambda^* \neq \lambda$. If $s^{2^n} \in Z$, $n = n(s)$, for all $s \in S$ then $D$ is commutative.

Proof. If $S \subset Z$ we are done; for, if $x \in D$ then $(\lambda + \lambda^*)x = \lambda^*(x + x^*) + (\lambda^* x^* + \lambda x)$ so is in $Z$. Since $\lambda + \lambda^* \neq 0$ we get that $x \in Z$.

So we may assume that $S \not\subset Z$; therefore $S$ generates $D$. By our hypothesis, there is an $a \in S$, $a \notin Z$ such that $a^2 \in Z$. Since $S$ generates $D$, there is a $b \in S$ such that $ab \neq ba$; again, using our hypothesis on $S$, we may assume that $b^2 a = ab^2$. Thus $0 \neq c = ab + ba$ commutes with both $a$ and $b$, hence $at + ta = 1$ where $t = c^{-1}b$ is symmetric. Now, $ta = 1 + at$ commutes with $at$; also

$(\lambda + \lambda^*)(at) = \lambda(at + ta) + \lambda ta + \lambda^* at = \lambda + \lambda ta + \lambda^* at$, and since

$\lambda ta + \lambda^* at \in S$, $((\lambda + \lambda^*)(at))^{2^n} = (\lambda + (\lambda ta + \lambda^* at))^{2^n} \in Z$ for suitable $n$.

This gives $(at)^{2^n} \in Z$. Therefore $(at)^{2^n} = (ta)^{2^n}$. But

$1 = (at + ta)^{2^n} = (at)^{2^n} + (ta)^{2^n} = 0$, a contradiction. This establishes the lemma.

LEMMA 3.2.3. If char $D = 2$ and every $a \in S$ satisfies $a^{2^n} \in Z$, $n = n(a)$, then $\dim_Z D \leq 4$ and all $x + x^*$ and $xx^*$ are in $Z$.

<u>Proof.</u> If $S \subset Z$, by Theorem 2.1.6 we have $\dim_Z D \leq 4$. So we may suppose that $S \not\subset Z$.

As in the proof of the preceding lemma we can find $a, t \in S$ with $a^2 \in Z$, $at^2 = t^2 a$ such that $at + ta = 1$. Thus $at$ commutes with $ta$ and $(at)^2 + at = a^2 t^2$. Since $t \in Z$, $t^{2^n} \in Z$, hence

$(at)^{2^n} + (at)^{2^{n-1}} = ((at)^2 + at)^{2^{n-1}} = a^{2^n} t^{2^n} = \beta \in Z$. If $c = (at)^{2^{n-1}}$ then we have $c^2 = c + \beta$, $\beta \in Z$ and since $c^* = (ta)^{2^{n-1}}$, $c + c^* = 1$.

Thus, if $xc = cx$ we have $x^* c^* = c^* x^*$ and so $x^* c = cx^*$ because $c^* = 1 + c$. Therefore, $C_D(c)^* = C_D(c)$. But $c \neq c^*$ is in $Z(C_D(c))$; by Lemma 3.2.2, $C_D(c)$ is commutative, hence must be a maximal subfield of $D$. By the double centralizer theorem [3], $Z(C_D(c)) = Z(c)$ and, since $C_D(c) = Z(C_D(c))$ we have that $Z(c)$ is a maximal subfield of $D$. However, $c$ is quadratic over $Z$, whence $[Z(c): Z] = 2$ and so $\dim_Z D \leq 4$.

To verify that $x+x^*$ and $xx^*$ are in $Z$ for all $x \in D$, knowing our conditions and that $\dim_Z D \leq 4$, is now trivial.

We are now able to prove a theorem due to Chacron [1] which generalizes the result of Kaplansky that we quoted earlier.

THEOREM 3.2.1. Let $D$ be a division ring in which $s^{n(s)} \in Z$ for all $s \in S$, where $n(s) \geq 1$. Then $\dim_Z D \leq 4$ and all $x+x^*$ and $xx^*$ are central in $D$. In particular, if char $D \neq 2$, then $S \subset Z$.

<u>Proof.</u> By Lemmas 3.2.1 and 3.2.3 we are done unless there is some element $s \in S$, $s \notin Z$ which is separable over $Z$.

If $Z^+ = Z \cap S$ and $L = Z^+(s)$, then every element $u \in L$ is symmetric; hence $u^{n(u)} \in Z^+$ for some $n(u) \geq 1$. By a result of Kaplansky [10], since $L \neq Z^+$ and $L$ is <u>not</u> purely inseparable over $Z^+$, $L$ must be algebraic over $GF(p)$ for some prime $p$. Hence $Z^+ \subset L$ is algebraic over $GF(p)$. Since $Z$ is quadratic over $Z^+$, $Z$ is algebraic over $GF(p)$; since $S$ is algebraic over $Z$, $S$ is algebraic over $GF(p)$. But then Theorem 3.1.1 tells us that $D$ is commutative. The theorem is now established.

For most of the theorems proved, once the result was established for certain conditions on the symmetric elements, we could prove a counterpart, under similar conditions on the skew elements. We do so again, now, proving a skew form of Theorem 3.2.1

THEOREM 3.2.2. Let $D$ be a division ring in which $k^{n(k)} \in Z$, $n(k) \geq 1$, for all $k \in K$. Then $\dim_Z D \leq 4$ and $k^2 \in Z$ for all $k \in K$.

<u>Proof.</u> If char $D = 2$, the present theorem is just a special case of Theorem 3.2.1. Hence we may assume, in what follows, that char $D \neq 2$.

If $0 \neq k \in K$ and $k^2 \notin Z$ then every element in $L = Z^+(k^2)$ is symmetric; hence, if $t \in L$, since $tk = kt$, $tk$ is skew and so $(tk)^n \in Z$ for some $n$. Because $k^m \in Z$ we get $t^{mn} \in Z$, and so $t^{mn} \in Z^+$. By the result of Kaplansky, either $L$ is purely inseparable over $Z^+$ or $L$ is algebraic over $GF(p)$. In the latter possibility, we would have $Z^+$, and so $Z$, algebraic over $GF(p)$; together with $K$ algebraic over $Z$ we would get that $K$ is algebraic over $GF(p)$. By Theorem 3.1.2 we would then be done. So we must assume that $k^2$ is purely inseparable over $Z$, whence $k^{2p^n} \in Z$ for some $n$, where $p = \text{char } D \neq 2$. The element $a = k^{p^n}$ is then skew, and $a^2 \in Z$.

Consider $D_1 = C_D(a)$. If $s \in D_1 \cap S$ then $sa$ is skew, hence $(sa)^q \in Z$; since $a^2 \in Z$ we see that $s^{2q} \in Z$. By Theorem 3.2.1 we conclude that all symmetric elements in $D_1$ are in $Z(D_1)$. Since $a \in Z(D_1)$ and $a$ is skew we have that $D_1$ is a field. Hence $C_D(a)$ is a maximal subfield. However, $a^2 \in Z$ and $C_D(a) = Z(a)$ is a maximal subfield of $D$, and is 2-dimensional over $Z$. The net result is that $\dim_Z D \leq 4$. From this point it is extremely easy to finish the proof that $k^2 \in Z$ for all $k \in K$.

We cannot make any further statement about the $D$ in Theorem 3.2.2, for if $D$ is any 4-dimensional division algebra in which $S \subset Z$ then $k^2 \in Z$ for all $k \in K$. There is no lack of such division algebras.

In chapter 6 we shall give some broad generalizations of these results.

We now turn to another type of commutativity hypothesis. A theorem of ours [4] states that if in the ring R, $a - a^2 p_a(a) \in Z$, for every $a \in R$, where $p_a(t)$ is a polynomial with integer coefficients depending on a, R must be commutative. Note that the condition merely says that $a + m_2 a^2 + \ldots + m_k a^k \in Z$ for suitable integers $m_2, \ldots, m_k$ which depend on a.

To avoid needless repetition, it is understood that the polynomials we shall be using for the next few pages will all have integer coefficients.

We want to study division rings with involution in which

1. $s - s^2 p_s(s) \in Z$ for all $s \in S$ (or perhaps, $t - t^2 p_t(t) \in Z$ for all $t \in T$),

or

2. $k - k^2 p_k(k) \in Z$ for all $k \in K$ (or perhaps, $k - k^2 p_k(k) \in Z$ for all $k \in K_0$).

The symmetric case was essentially done in [7]. The skew case was handled by Lee [12]. Before proving the theorems about this situation, we must dispose of the special case in which $x = x^2 p_x(x)$ for all $x \in S$ or for all $x \in K$. We do this in

THEOREM 3.2.3. Suppose that in D

1. $t = t^2 p_t(t)$ for all $t \in T$,

or

2. $k = k^2 p_k(k)$ for all $k \in K_0$.

Then D is commutative. Moreover, if $K_0 \neq 0$ (i.e., if * is not the identity map), then D is algebraic over GF(p) for some prime p.

Proof. If $K_0 = 0$ then D, of course, must be commutative. Even in this case, if char $D \neq 2$, in the presence of Condition 1, D must be algebraic over GF(p). We make the argument for the case of $K_0$; a similar argument works for T. So we assume that $K_0 \neq 0$ and that $k - k^2 p_k(k) \in Z$ for all $k \in K_0$.

It is enough to show that char $D \neq 0$; for, if so, then $K_0$ is algebraic over GF(p) hence the result is a consequence of Theorem 3.1.2. Let $a \neq 0 \in K_0$; thus $a = a^2 p(a)$, so $ap(a) = 1$. This tells us that $a^{-1}$ is integral over the ring of integers. If char $D = 2$ we are done. If char $D \neq 2$ then $2a \neq 0$ is also in $K_0$, so $2a^{-1} = a^{-1}(2a)a^{-1} \in K_0$, hence $(2a^{-1})^{-1} = 2^{-1}a$ is integral over the integers. Therefore $(2^{-1}a)(a^{-1})$ is integral over the integers, which is to say, $2^{-1}$ is integral over the integers. This clearly implies that char $D \neq 0$, thereby proving the theorem.

We now proceed to

THEOREM 3.2.4. Suppose that in D, $s - s^2 p_s(s) \in Z$ for every $s \in S$. Then $S \subset Z$ and $\dim_Z D \leq 4$.

Proof. S is, by hypothesis, algebraic over Z. Suppose that there is an $s \in S$, $s \notin Z$ which is separable over Z. If $Z^+ = Z \cap S$ and $L = Z^+(s)$ then $L \neq Z^+$; moreover, since every element $u \in L$ is symmetric, $u - u^2 p_u(u) \in Z^+$. By a result of ours [4] on fields either L is purely inseparable over $Z^+$ — which it isn't since $s \in L$ is separable over $Z^+$ — or L, and so $Z^+$, is algebraic over GF(p) for some prime p. Since Z is quadratic over $Z^+$ we have that Z is algebraic over GF(p). Finally, since S is algebraic over Z, S must be algebraic over GF(p), hence every element $s \in S$ is periodic. By

Theorem 3.1.1, D must be commutative, contradicting that $s \notin Z$. Hence we have that all $s \in S$, $s \notin Z$ are not separable over Z.

Suppose that $S \not\subset Z$, and let $s \in S$, $s \notin Z$. By the above, s is not separable over Z. Thus char $D = p \neq 0$. Now $\gamma = s - s^2 p_s(s) \in Z$, so that s is a root of $q(x) = x - x^2 p_s(x) - \gamma$ over Z. Therefore q(x) has a multiple root, and so q(x) and q'(x) have a common root, a, in some extension of Z. But $0 = q'(a) = 1 - a^2 p_s'(a) - 2ap_s(a)$, whence a is algebraic over GF(p). However, $0 = q(a) = a - a^2 p_s(a) + \gamma$ from which we see that $\gamma$ is algebraic over GF(p). But then s is algebraic over GF(p) since it is a root of $q(x) = x - x^2 p_s(x) + \gamma$. But this forces s to be separable over Z, and so in Z. With this contradiction the theorem is established.

It is a fairly easy exercise to show that if D is a division algebra of characteristic 2 in which all the symmetric elements are in Z (hence $\dim_Z D \leq 4$) then D must, in fact, be commutative. Therefore we have

COROLLARY. If char $D = 2$ and $s + s^2 p_s(s) \in Z$ for every $s \in S$, then D is commutative.

For the skew elements we have a sharper version of Theorem 3.2.4. This result is due to P. H. Lee [12].

THEOREM 3.2.5. Suppose that in D every $k \in K$ satisfies $k - k^2 p_k(k) \in Z$. Then D is commutative.

Proof. If char $D = 2$ this result is exactly that expressed in the Corollary to Theorem 3.2.4. Hence we may assume that char $D \neq 2$.

If $Z \cap K = 0$, since $k - k^2 p(k) \in Z$, applying * we get $-k - k^2 p(-k) \in Z$, whence $b = 2k - k^2(p(k) - p(-k)) \in Z$. But $b \in K$;

hence $b = 0$. This yields $2k = k^2(p(k) - p(-k))$, and so $k = k^2q(k)$ where $q$ is a polynomial with integer coefficients which depend on $k$. By Theorem 3.2.3 we conclude that $D$ is commutative.

Se we must suppose that $Z \cap K \neq 0$. Let $\gamma \neq 0 \in Z \cap K$; then $\gamma^* = -\gamma$ and $S = \gamma K$, $K = \gamma S$. If $s \in S$ then $\gamma s \in K$ hence $\gamma s - \gamma^2 s^2 p(\gamma s) \in Z$, and so $s - \gamma s^2 p(\gamma s) \in Z$. In other words, for some integers $m_i$, $t = s + \gamma m_1 s^2 + \ldots + \in Z$; therefore $t^* = s - \gamma m_1 s^2 + \gamma^2 m_2 s^3 - \ldots \in Z$. This gives, since $\operatorname{char} D \neq 2$, that $w = s + \gamma^2 m_2 s^3 + \gamma^4 m_4 s^5 + \ldots + \gamma^{2k} m_{2k} s^{2k+1} \in Z$. Thus $S$ is algebraic over $Z$. If $S \not\subset Z$ then, since $\operatorname{char} D \neq 2$, there is an element $s \in S$, $s \notin Z$ which is algebraic over $Z$. If $Z$ were algebraic over a finite field then every element in $S$ would be periodic; from this we would have, by Theorem 3.1.1, that $D$ is commutative. We may therefore assume that $Z$ is not algebraic over a finite field.

The reduction we have carried out allows us to invoke a result on valuation theory due to Nagata, Nakayama and Tuzuku [15] to the pair of fields $Z(s)$ and $Z$. By a simple extension of their result there exists a logarithmic valuation $v$ on $Z$ (in fact, there exists an infinity of such valuations) where $v(m_2) = 0, \ldots, v(m_{2k}) = 0$, $v(\gamma^2) = 0$ which has two distinct extensions, $V_1$ and $V_2$, to $Z(s)$, $V_1(s) \neq V_2(s)$. We may suppose that $V_1(s) < 0$, otherwise we would carry out the argument on $s^{-1}$. Thus $V_1(\gamma^{2i} m_{2i} s^{2i+1}) = (2i+1)V_1(s) < V_1(s)$ for $i \neq 0$. Since the valuation $V_1$ is non-archimedean we have that

$$\begin{aligned} V_1(w) &= V_1(s + \gamma^2 m_2 s^3 + \ldots + \gamma^{2k} m_{2k} s^{2k+1}) \\ &= \min(V_1(s), \ldots, (2i+1)V_1(s), \ldots, (2k+1)V_1(s)) \\ &= (2k+1)V_1(s) . \end{aligned}$$

We claim that $V_2(s) < 0$. For, if $V_2(s) \geq 0$ then

$$V_2(w) = V_2(s + \gamma^2 m_2 s^3 + \ldots + \gamma^{2k} m_{2k} s^{2k+1})$$
$$\geq \min(V_2(s), \ldots, (2k+1)V_2(s)) \geq 0 .$$

But since $V_1, V_2$ are extension of v, and since $w \in Z$,

$$0 \leq V_2(w) = v(w) = V_1(w) = (2k+1)V_1(s) < 0 ,$$

a contradiction. In consequence, $V_2(s) < 0$. But then the same computation as that made for $V_1$ shows that $V_2(w) = (2k+1)V_2(s)$. Thus $(2k+1)V_2(s) = V_2(w) = v(w) = V_1(w) = (2k+1)V_1(s)$, leading us to the contradiction $V_1(s) = V_2(s)$. With this we conclude that $S \subset Z$.

But $K = \gamma S$, $\gamma \in Z$, whence $K \subset Z$. Since $D = S + K$ we end up with the fact that D is commutative.

## 3. Rings with Periodic Skew or Symmetric Elements

In the first section of this chapter we saw that division rings in which $s^{n(s)} = s$ for all $s \in S$ turned out to be fields. A similar result held for the skew elements. We would like to study more general rings in which such a periodicity condition holds for the symmetric elements or for the skew elements. However, no matter how decent such a ring may be, we have little hope of proving it to be commutative. For instance, if F is a field algebraic over a finite field of characteristic $p \neq 2$, let $R = F_2$, the ring of $2 \times 2$ matrices over F, and let * be the symplectic involution on $F_2$. Then it is immediate that $s^{n(s)} = s$ for every $s \in S$. The ring R is certainly not commutative, yet it is of a very easy structure — a finite dimensional simple algebra over F.

Similarly for the same $R$ as above, using transpose as the involution on $F_2$, $k^{n(k)} = k$ for every $k \in K$.

What we shall show is that fields and $2 \times 2$ matrices over fields provide us with the basic building blocks for rings in which a periodicity condition is imposed on the symmetric elements or on the skew elements. The result in the symmetric case was obtained by Montgomery [13], [14]. She made use of a result of Osborn (who had exploited Theorem 3.1.1 to get his result) about Jordan rings. However, she left the skew case open. The proof we give here avoids the use of results about Jordan algebras and treates the skew and symmetric cases at the same time. The treatment we give comes from [5].

We begin with an easy result from linear algebra.

LEMMA 3.3.1. Let $F$ be a field which is algebraic over a finite field. Suppose that in $F_n$, the ring of $n \times n$ matrices over $F$, there is an involution $*$ such that $xx^* \neq 0$ if $x \neq 0$. Then $n = 1$ or $2$. Moreover, if char $F = 2$ then $n = 2$ is not possible.

Proof. Since $F$ is the center of $F_n$, we must have $F^* \subset F$. Thus $*$ induces an automorphism $\bar{\ }$ on $F$. Let $F_0$ be the fixed field of $\bar{\ }$, that is, $F_0 = \{\alpha \in F \mid \bar{\alpha} = \alpha\}$.

If $e_{ij}$ are the usual matrix units, then, since $e_{ii}e_{ii}^* \neq 0$, it follows easily that we may assume that $e_{ii}^* = e_{ii}$ for $i = 1, 2, \ldots, n$. Since $e_{ij} = e_{ii}e_{ij}e_{jj}$, applying $*$ and using $e_{ii}^* = e_{ii}$, $e_{jj}^* = e_{jj}$ we obtain that $e_{ij}^* = \alpha_{ij}e_{ji}$ where $\alpha_{ij} \in F_0$ and where $\alpha_{ij}^{-1} = \alpha_{ji}$.

Since $F_0$ is algebraic over a finite field, we can find three elements, $a_1, a_2, a_3$, not all 0 in $F_0$ such that $a_1^2 + \alpha_{12}a_2^2 + \alpha_{13}a_3^2 = 0$.

If char F = 2 we can, in fact, find $a_1, a_2$ in $F_0$ such that $a_1^2 + \alpha_{12}a_2^2 = 0$.

Suppose that $n \geq 3$. Consider the matrix

$$0 \neq x = \begin{pmatrix} a_1 & a_2 & a_3 & 0 & \dots & 0 \\ 0 & 0 & 0 & 0 & \dots & 0 \\ \vdots & \vdots & \vdots & \vdots & & \vdots \\ 0 & 0 & 0 & 0 & \dots & 0 \end{pmatrix}$$

in $F_n$. Thus

$$x^* = \begin{pmatrix} a_1 & 0 & \dots & 0 \\ \alpha_{12}a_2 & 0 & \dots & 0 \\ \alpha_{13}a_3 & 0 & \dots & 0 \\ 0 & 0 & \dots & 0 \\ \vdots & \vdots & & \vdots \\ 0 & 0 & \dots & 0 \end{pmatrix}$$

and $xx^* = (a_1^2 + \alpha_{12}a_2^2 + \alpha_{13}a_3^2)e_{11} = 0$.

Thus, by our assumption on $*$, we must have $n \leq 2$. If char F = 2 and n = 2, using the matrix $x = \begin{pmatrix} a_1 & a_2 \\ 0 & 0 \end{pmatrix}$, where $a_1^2 + \alpha_{12}a_2^2 = 0$, we have $xx^* = 0$ yet $x \neq 0$. So, if char F = 2, n must, in fact, be 1. The lemma is now proved.

We can now pass to finite rings with the property $xx^* \neq 0$ if $x \neq 0$.

LEMMA 3.3.2. Let R be a finite ring with involution $*$ in which $xx^* \neq 0$ if $x \neq 0$. Then R is the direct sum of finite fields and rings of $2 \times 2$ matrices over finite fields. These $2 \times 2$ matrix rings must be over fields of characteristic not 2.

Proof. R clearly has no nilpotent symmetric elements. Thus R is semi-simple. For if N is the radical of R and $x \in N$ then $xx^* \in N$

so must be nilpotent since R is finite. But $xx^*$ is symmetric; hence $xx^* = 0$ whence $x = 0$.

Since R is semi-simple and finite it is a direct sum of finite fields and total matrix rings over finite fields. We write it as $R = R_1 \oplus \ldots \oplus R_k$, where the $R_i$ are simple, and are the minimal ideals of R. Since $xx^* \neq 0$ if $x \neq 0$, we have $R_i R_i^* \neq 0$; this gives us $R_i^* = R_i$. Hence the simple components of R are invariant re *.

If some $R_i = F_n$, then by Lemma 3.3.1 we have $n = 1$ or 2, and if char $F = 2$, $n = 2$ is not possible. Thus the lemma is proved.

Note that if char $R = 2$ in this last lemma, R must be commutative. Also, if R is <u>not</u> commutative, some $2 \times 2$ matrix ring must be a constituent of R.

We shall be studying rings with various periodicity conditions. To avoid a lengthy repetition of the condition each time, we give the conditions that will come up numbers, and we shall refer to these by their numbers.

<u>Condition I:</u> $s^{n(s)} = s$, $n(s) > 1$, for all $s \in S$.

<u>Condition I':</u> $t^{n(t)} = t$, $n(t) > 1$, for all $t \in T$.

<u>Condition II:</u> $k^{n(k)} = k$, $n(k) > 1$, for all $k \in K$.

<u>Condition II':</u> $k^{n(k)} = k$, $n(k) > 1$, for all $k \in K_0$.

We prove

LEMMA 3.3.3. Let R be a semi-prime ring in which all the symmetric idempotents are central. Suppose that R satisfies Condition I' or that R satisfies Condition II'. Then R is a subdirect product of commutative rings and $2 \times 2$ matrices over fields.

Proof. Let P be a prime ideal of R. If $P^* \neq P$ then in $\overline{R} = R/P$, the non-zero ideal $\overline{P}^* = \frac{P + P^*}{P}$ is commutative. For, every element $\overline{x} \in \overline{P}^*$ can be realized as $\overline{x \pm x^*}$ where $x^* \in P$, so according as the condition is I' or II', since $\overline{x+x^*}^n = \overline{x + x^*} = \overline{x}$ or $\overline{x}^n = \overline{x - x^*}^n = \overline{x - x^*} = \overline{x}$, by Jacobson's theorem $\overline{P}^*$ is commutative. Since R is prime and $\overline{P}^* \neq 0$ is a commutative ideal of R, R is commutative.

If, on the other hand, $P^* = P$ then $\overline{R} = R/P$ has an involution induced by * and $\overline{R}$ inherits Condition I' or II' from R. If $\overline{K}_0 = 0$ then $\overline{R}$ is commutative; if $\overline{T} = 0$ then, if char $\overline{R} \neq 2$, we get $\overline{x} = -\overline{x}$ for all $\overline{x} \in \overline{R}$ and so $2\overline{x}^2 = 0$, whence $\overline{x}^2 = 0$. This is not possible in a semi-prime ring. Hence we may assume that $\overline{K}_0 \neq 0$ and $\overline{T} \neq 0$. If $0 \neq \overline{x} \in \overline{T}$ or $0 \neq \overline{x} \in \overline{K}_0$ it is an image of an element in T or $K_0$ respectively. But then $x^n = x$, $n > 1$, (depending on whether it is Condition I' or II') hence $e = x^{n-1}$ is an idempotent. It is clearly symmetric (since $e^2 = e$, and $e^* = \pm e$, we have $e^* = e$). So, by hypothesis, $e \in Z(R)$. Thus $\overline{e}$ is a central idempotent in $\overline{R}$; since $\overline{R}$ is prime, $\overline{e} = 0$ or 1. Because $0 \neq \overline{x} = \overline{x}\,\overline{e}$, we have $\overline{e} \neq 0$. Thus $\overline{e} = 1$; hence $\overline{x}^{n-1} = 1$. Therefore every non-zero element in $\overline{T}$ or $\overline{K}_0$ is invertible in $\overline{R}$. By Theorem 2.1.7 or 2.3.4, $\overline{R}$ is either a division ring or the ring of $2 \times 2$ matrices over a field. If $\overline{R}$ is a division ring, by Theorems 3.1.1 and 3.1.2 it is commutative.

Since R is semi-prime, $\bigcap P = 0$ where this intersection runs over all prime ideals of R, hence R is a subdirect product of the R/P's. Each R/P, we saw, must be commutative or the $2 \times 2$ matrices over a field. With this the lemma is proved.

We come to the key result in these considerations. Unfortunately, the proof of this result is long and difficult.

THEOREM 3.3.1. Let $R$ be a primitive ring satisfying Condition I' or satisfying Condition II'. Then $R$ is a field or the ring of all $2 \times 2$ matrices over a field. Moreover, if Condition II' holds then any two elements in $K_0$ must commute.

Proof. If Condition I' holds there are no non-zero nilpotent traces; if Condition II' holds there are no non-zero nilpotent skew-traces. Thus, in either case, by Theorem 2.2.4 and its Corollary, either $R$ is an order in $F_2$, or $xx^* = 0$ in $R$ implies that $x = 0$.

If $R$ is an order in $F_2$ then $R$ satisfies the standard identity of degree 4 (or, if you want, the identities of $F_2$). By Kaplansky's theorem $R$ must be a 4-dimensional simple algebra over its center. Hence either $R$ is a division ring or the $2 \times 2$ matrices over a field. If $R$ is a division ring it is commutative by Theorems 3.1.1 and 3.1.2. So, since $R$ is an order in $F_2$, we, in fact, get that $R$ must be the ring of $2 \times 2$ matrices over a field.

Thus, from now on, we may assume that $xx^* = 0$ in $R$ implies $x = 0$. As usual, we may also assume that $K_0 \neq 0$ and that $T \neq 0$. As we have seen, from the periodicity condition on $T$ or $K_0$, we conclude that the characteristic of $R$ is $p$, $p \neq 0$.

If $a \neq 0 \in T$ and $a^n = a$ then $e = a^{n-1} \neq 0$ is a symmetric idempotent. If $a \neq 0 \in K_0$ and $a^n = a$ then, here too, $e = a^{n-1}$ is a symmetric idempotent. Thus, in either Condition I' or II', $R$ must have a non-zero symmetric idempotent. If $1$ is the only symmetric idempotent, then Lemma 3.3.3 applies and $R$ is either a field or the $2 \times 2$ matrices over a field, since it is primitive. So we may suppose that in R there is a symmetric idempotent $e \neq 0, 1$.

If $e$ commutes with all symmetric elements in $R$ then $e$ centralizes $\overline{S}$. By Theorem 2.1.5, either $S \subset Z$ or $\overline{S}$ contains a non-zero ideal $U$ of $R$. If $S \subset Z$ then $e \in S \subset Z$; however in a prime ring (hence, also a primitive ring) any central idempotent must be 0 or 1. On the other hand, if $\overline{S} \supset U \neq 0$, $U$ an ideal of $R$, then $e$ centralizes $U$; by Lemma 1.1.6, $e \in Z$, which is again impossible. The upshot of this discussion is that there is a $b \in S$ with $c = eb - be \neq 0$. Clearly $c \in K_0$.

If we are operating in Condition II' then $c^n = c$ for some $n > 1$. On the other hand, if our running hypothesis is Condition I' but not II we claim that $c^n = c$, again, for some $n > 1$. Why? We may suppose that Condition II' does not hold, otherwise our claim would be automatically valid; thus char $R \neq 2$. But then $2c^2 \in T$ and we can find an integer $k > 1$ such that both $2^k \equiv 2 \bmod (\text{char } R)$ and $(2c^2)^k = 2c^2$. These yield $2(c^{2k} - c^2) = 0$ and so $c^{2k} - c^2 = 0$. This last relation gives us $(c^{2k-1} - c)^2 = 0$, hence $(c^{2k-1} - c)(c^{2k-1} - c)^* = 0$. Because $*$ is positive definite on $R$ we arrive at $c^{2k-1} = c$, $2k-1 > 1$, thus verifying our claim that $c$ is periodic.

Since $c = eb - be$ and $e^2 = e$ we have that $c = ec + ce$, from which we see that $c^2e = ec^2$. Because of these above relations the ring $A_0 = \{q_1(c) + q_2(c)e \mid q_1, q_2$ polynomial over $P = GF(p)\}$ is finite. Moreover, since $e^* = e$, $c^* = -c$, $A_0$ must be invariant re $*$; because $xx^* \neq 0$ for $x \neq 0$ in $R$, the same holds true for $A_0$. Also, $A_0$ is not commutative; if $A_0$ were commutative we would have $c = ec + ce = 2ec$, hence $ec = 2e^2c = 2ec$ and so $ec = 0$, $c = 2ec = 0$ a contradiction.

Therefore all the hypotheses of Lemma 3.3.2 are satisfied by $A_0$, in consequence of which $A_0$ is the direct sum of finite fields and $2 \times 2$ matrices over finite fields, where these direct summands are invariant with respect to $*$. Moreover, since $A_0$ is not commutative, one of these direct summands — say B — is a $2 \times 2$ matrix ring over a finite field F of characteristic not 2.

We look at $*$ on $B = F_2$; as in the proof of Lemma 3.3.1 we may assume that we have matrix units $e_{ij}$ in B such that $e_{11}^* = e_{11}$, $e_{22}^* = e_{22}$, $e_{12}^* = \alpha e_{21}$ and $e_{21}^* = \alpha^{-1} e_{12}$ where $\alpha^* = \alpha$ is in F. Thus $\alpha^m = 1$ for some $m > 0$ ($m = p^r - 1$ will do, for some r) where m is invertible modulo p.

Since B is a direct summand of the semi-simple ring $A_0$, $B = fA_0$ where $f^* = f = f^2 \neq 0$. Therefore the ring $R_0 = fRf$ is a primitive ring invariant with respect to $*$. Moreover its $K_0$ and its T lie, respectively in $K_0 \cap R_0$ and $T \cap R_0$, whence $R_0$ inherits the hypothesis of R. Our objective now is to show that $R_0$ must be the ring of all $2 \times 2$ matrices over a field.

Consider $C_{R_0}(\alpha)$; certainly $C_{R_0}(\alpha)$ contains the field $P(\alpha) \subset F$ and is an algebra over $P(\alpha)$. Moreover $C_{R_0}(\alpha)$ contains the $2 \times 2$ matrices over $P(\alpha)$. Therefore $C_{R_0}(\alpha) = (P(\alpha))_2 \otimes G$ where G is the centralizer in $C_{R_0}(\alpha)$ of $(P(\alpha))_2$. Since $xx^* \neq 0$ for $x \neq 0$ in G, G must be semi-prime.

If in G, every symmetric idempotent commutes with every symmetric element of G then every such idempotent must be in the center of G. For, if $e^* = e = e^2 \in G$ and $d(x) = xe - ex$ then from $d(S \cap G) = 0$ and the fact that char $R_0 \neq 2$ we get $d^2(k) = 0$ for $k \in K \cap G$, and so $d^2(x) = 0$ for all $x \in G$. But, because $e^2 = e$, $d^3(x) = d(x)$, in consequence of which, $d(x) = 0$ for all $x \in G$. Thus $e \in Z(G)$. By Lemma 3.3.3, G satisfies a polynomial identity. Since G is semi-prime we have that $C_{R_0}(\alpha) = (P(\alpha))_2 \otimes G$ also satisfies a polynomial identity, hence by Montgomery's theorem (Theorem 1.6.2), $R_0$ satisfies a polynomial identity. Since $R_0$ is primitive, by Kaplansky's theorem $R_0$ must be simple artinian.

On the other hand, if there exists a symmetric idempotent e in G and a symmetric element $t \in G$ such that $et \neq te$, using the argument used earlier for R we can produce a $2 \times 2$ matrix ring in G which is invariant with respect to *. So, in G there would be a self-adjoint idempotent g such that gGg contains the $2 \times 2$ matrices over a field $L \supset P(\alpha)$, which is algebraic over a finite field, and in which $L_2^* = L_2$. Thus $gC_{R_0}(\alpha)g = (P(\alpha))_2 \otimes L_2 = L_4$; but $xx^* \neq 0$ for $x \neq 0$ in R, hence in $L_4$. Since L is algebraic over a finite field, by Lemma 3.3.1 this is not possible. Hence this second possibility does not arise, and so $R_0$ must be simple artinian.

Let $h^2 = h \in R_0$ be such that $hR_0$ is a minimal right ideal of $R_0$; thus $hR_0h$ is a division ring. But $hR_0h = hfRfh = hRh$, since $fh = hf = h$, thus hRh is a division ring. Consequently hR is a minimal right ideal of R. By the general structure theorem for primitive rings with involution having a minimal right ideal (Theorem 1.2.2) and since $xx^* \neq 0$ for $x \neq 0$ in R, if R were not the ring of $2 \times 2$ matrices over a field, R would contain a subring, invariant re *, isomorphic to the $3 \times 3$ matrices over a division ring D. D satisfies Condition I' or II' so by Theorem 3.1.2 and its Corollary 2, D is a field and is algebraic over a finite field. In $D_3$, $xx^* \neq 0$ for $x \neq 0$; hence by Lemma 3.3.1 this is not possible. The net result of all this is that $R$ must be $F_2$ for some field $F$. Thus the larger part of the theorem is proved.

What remains is to show that if Condition II' holds, that is, if $(x - x^*)^{n(x)} = x - x^*$ for all $x \in R = F_2$ then any two elements in $K_0$ commute.

If the involution on $F_2$ is symplectic then, if char $F = 2$, $x + x^*$ is a scalar, so certainly any two skew-traces commute. If char $F \neq 2$ then $\begin{pmatrix} 0 & 1 \\ 0 & 0 \end{pmatrix}$ is a skew-trace and is nilpotent, so cannot be periodic.

Therefore we may assume that $*$ on $F_2$ is of transpose type, that is to say,

$$\begin{pmatrix} a & b \\ c & d \end{pmatrix} = \begin{pmatrix} \bar{a} & \alpha^{-1}\bar{c} \\ \alpha\bar{b} & \bar{d} \end{pmatrix}$$

where $\bar{\ }$ indicates the automorphism induced by $*$ on $F$, and where $\alpha^* = \bar{\alpha} = \alpha$ is in $F$.

If $\bar{a} = a$ for all $a \in F$ then any skew trace is of the form $\begin{pmatrix} 0 & b \\ -\alpha b & 0 \end{pmatrix}$ and any two such commute. Suppose then that $\bar{a} \neq a$ for some $a \in F$, and let $b = a - \bar{a} \neq 0$. The field $L = P(\alpha, b)$ is finite since $\alpha, b$ are algebraic over $P$, and $L^* = L$ since $\alpha^* = \alpha$, $b^* = -b$. Let $L_0 = \{x \in L \mid \bar{x} = x\}$; if char $P \neq 2$ then $[L: L_0] = 2$, and since $L$ is finite, every element in $L_0$ is a norm in $L$, that is, is of the form $u\bar{u}$ for some $u \in L$. If char $P = 2$, every element in $L$ is a square and $L = L_0$ hence, here too, every element in $L_0$ is a norm. Thus, in either case, $-\alpha^{-1} = \beta\bar{\beta}$ for some $\beta \in L$. The matrix

$$x = \begin{pmatrix} b & b\beta \\ \alpha b \bar{\beta} & -b \end{pmatrix}$$

is then a skew trace, and as is easily verified, $x^2 = 0$. This contradicts the periodicity of the skew-traces in $F_2$. The theorem is finally proved.

Theorem 3.3.1 represented the major hurdle to the development of the various theorems about rings in which some periodicity condition holds. With it out of the way, the road is clear to obtain a series of interesting theorems. The next three theorems are due to Montgomery [13, 14].

THEOREM 3.3.2. Let $R$ be a semi-simple ring with involution $*$ such that $(x+x^*)^{n(x)} = x+x^*$, $n(x) > 1$, for every $x \in R$. Then $R$ is a subdirect product of fields and $2 \times 2$ matrix rings over fields. In consequence $R$ satisfies the standard identity in 4 variables.

Proof. Let $P$ be a primitive ideal of $R$. If $P^* \neq P$ then $A = (P + P^*)/P$ is a non-zero ideal in the primitive ring $R/P$. Moreover, every element in $A$ is the image of an element of the form $u+u^*$ where $u \in P^*$ (for if $a \in A$, $a = x_1 + x_2^* + P = x_2^* + P$ where $x_2 \in P$ and so $a = x_2 + x_2^* + P$). Therefore if $a \in A$ then $a^{n(a)} = a$, whence, by Jacobson's theorem, $A$ is commutative. Since $R/P$ is a primitive ring containing a non-zero commutative ideal $A$, $R/P$ must be commutative, and so a field.

If, on the other hand, $P^* = P$ then $R/P$ is a primitive ring with $*$ and inherits the property of the periodicity of all $x+x^*$. By Theorem 3.3.1 $R/P$ is either a field or the $2 \times 2$ matrices over a field.

Since $R$ is semi-simple it is the subdirect product of the $R/P$'s, hence, by the above, is the subdirect product of fields and $2 \times 2$ matrix rings over fields. Because these satisfy the standard identity in 4 variables, so must $R$. This finishes the proof of the theorem.

The hypothesis of Theorem 3.3.2 is on T, the set of traces of R. If we impose the same hypothesis on S itself we can prove a stronger theorem. This is

THEOREM 3.3.3. Let R be a semi-prime ring with involution such that $s^{n(s)} = s$, $n(x) > 1$, for every $s \in S$. Then R is a subdirect product of fields and $2 \times 2$ matrices over fields; in consequence, R satisfies the standard identity in 4 variables. If char R = 2 then R must be commutative, even if R is not semi-prime.

Proof. We first show that R is semi-simple. If J is the radical of R and $a^* = a \in J$, then, because $a^n = a$, we must have $a = 0$. Thus $J \cap S = 0$. Since $J^* = J$ we must have that every element in J must be skew. In particular, if char R = 2, J = 0. If $u \in J$ then, since u is skew, $u^2$ is symmetric hence $u^2 = 0$. Now, for $x \in R$, $ux \in J$ hence $(ux)^* = -ux$; thus $0 = x^* u^2 = (ux^*)u = -uxu$. In short, uR is a nilpotent right ideal of R. Since R is semi-prime we have that $u = 0$, whence $J = 0$ and R is semi-simple. By Theorem 3.3.2 we have the subdirect product structure of R as asserted in the theorem.

We still must show that if char R = 2 then R is commutative. This trivially reduces to showing that if char F = 2 and F is finite then in $F_2$ it is not possible that $s^{n(s)} = s$ for every symmetric element. The proof at the end of that of Theorem 3.3.1 shows that we may assume that the involution is of transpose type and * on F is the identity map. Hence $\begin{pmatrix} a & b \\ c & d \end{pmatrix}^* = \begin{pmatrix} a & \alpha^{-1}c \\ \alpha b & d \end{pmatrix}$ for some $\alpha \in F$. Since F is finite of characteristic 2, $\alpha = \beta^2$ for some $\beta \in F$. The matrix

$\begin{pmatrix} \beta & 1 \\ \alpha & \beta \end{pmatrix}$ is then symmetric and nilpotent, hence not periodic. The proof is now complete.

Even when R is not semi-prime we can say a fair amount about its structure. This is the content of

THEOREM 3.3.4. Let R be a ring with involution in which $s^{n(s)} = s$, $n(s) > 1$, for every $s \in S$. If J is the radical of R then $J^3 = 0$ and $J[R, R] = 0$. The ring R/J is a subdirect product of fields and 2 × 2 matrix rings over fields. Finally, R satisfies $S_4(x_1, \ldots, x_4)^2$ and $S_6(x_1, \ldots, x_6)$ where $S_n$ is the standard identity of degree n.

<u>Proof.</u> As in the proof of the preceding theorem, $J \cap S = 0$ and $a^* = -a$, $a^2 = 0$ for every $a \in J$. Since $a^* \neq a$ and $a^* = -a$ for any $a \neq 0 \in J$, J must be 2-torsion free. From this we have that $J^3 = 0$.

If $a \in J$, $x \in R$ then $(ax)^* = -ax$, that is, $x^*a = ax$. Therefore $(xy)^*a = axy$; however $(xy)^*a = y^*x^*a = y^*ax = ayx$. These last relations give $a(xy - yx) = 0$, and so $J[R, R] = 0$.

Since R/J inherits the basic hypothesis we have imposed on S in R, by Theorem 3.3.2 we have the structure of R/J as stated in the assertion of Theorem 3.3.4. Thus, if $a_1, a_2, a_3, a_4 \in R$ then $S_4(a_1, \ldots, a_4) \in J$ and so has square 0. Also, since $S_4(a_1, \ldots, a_4)(xy - yx) \in J(R, R] = 0$, we trivially get that R satisfies $S_6(x_1, \ldots, x_6)$.

We close this section by deriving the results just obtained for the symmetric elements for the case of the skew elements. Here, however, there is something new that we can say that we could not say before. This new fact is that under the hypothesis imposed it will turn out that any two skew elements must commute. This is not true, for the preceding theorems, for the symmetric elements. For instance, if $R = F_2$ where $F$ is the field of three elements, using transpose as the $*$ it is easy to verify that $s^{n(s)} = s$ for every symmetric element; yet the symmetric elements $\begin{pmatrix} 1 & 0 \\ 0 & -1 \end{pmatrix}$ and $\begin{pmatrix} 0 & 1 \\ 1 & 0 \end{pmatrix}$, say, do not commute.

The result that under the imposed periodicity condition skew elements must commute verifies a question raised by Jacobson [10] for a particular case. Jacobson asked whether a restricted Lie algebra in which the restriction map is periodic on each element in the sense that $a^{p^{n(a)}} = a$ must be trivial. For the case when this Lie algebra is the set of skew elements of an associative ring with involution we shall see that the answer is yes.

The results come from the paper [5].

THEOREM 3.3.5. Let $R$ be a semi-simple ring in which $k^{n(k)} = k$, $n(k) > 1$, for every $k \in K$. Then $R$ is a subdirect product of fields and $2 \times 2$ matrices over fields. Moreover, if $a, b \in K$ then $ab = ba$.

Proof. The proof of the theorem is like that of Theorem 3.3.2. If $P$ is a primitive ideal of $R$ and $P^* \neq P$ then, as in the proof of Theorem 3.3.2, $R/P$ must be commutative. So, here, $ab - ba \in P$ for all $a, b \in R$, hence certainly for all $a, b \in K$. On the other hand, if $P^* = P$ and if $R/P$ is not of characteristic 2, then $R/P$ inherits the hypothesis on $K$ so by Theorem 3.3.1, $R/P$ is a field or a $2 \times 2$ matrix ring over a field and all skew elements in $R/P$ commute. So, if $a, b \in K$ then $ab - ba \in P$. If $R/P$ is of characteristic 2 then $2(ab - ba) \in P$ for all $a, b \in K$ (and $R/P$ is a field or $2 \times 2$ matrices). Hence $2(ab - ba)$

is in every primitive ideal, for every $a, b \in K$. Because $R$ is semi-simple, we have that $2(ab - ba) = 0$.

If $a \in K$ then $a^n = a$ and $(2a)^n = 2a$ for some $n > 1$, hence $2(2^{n-1} - 1)a = 0$. Thus, if $M = \{x \in R \mid 2x = 0\}$ then $(2^{n-1} - 1)a \in M$. But $M$ is of characteristic 0; by Theorem 3.3.3, $M$ is commutative. Hence, if $a, b \in K$ then $(2^{n-1} - 1)a \in M$ and $(2^{m-1} - 1)b \in M$ whence these two elements commute. Using that $2(ab - ba) = 0$ we derive from these relations that $ab = ba$.

The theorem is now proved.

We can sharpen the last part of Theorem 3.3.5 to hold in any ring. This is

THEOREM 3.3.6. Let $R$ be a ring with involution in which $k^{n(k)} = k$, $n(k) > 1$, for every $k \in K$. Then any two elements of $K$ must commute.

Proof. If $K$ is 2-torsion free it is easy to see that $R/J$ inherits the property of periodicity we have imposed on $K$. By Theorem 3.3.5 we then have $ab - ba \in J$ for any $a, b \in K$. Since $(ab - ba)^n = (ab - ba)$ and $ab - ba \in J$ we must have $ab - ba = 0$, and so any two elements of $K$ commute.

Let $M = \{x \in R \mid 2x = 0\}$. As in the proof of Theorem 3.3.5, $(2^{n-1} - 1)a \in M$ for $a \in K$ with $n > 1$. From this we easily see that the skew elements in $R/M$ are 2-torsion free, hence by the paragraph above, any two skew elements in $R/M$ commute. Therefore, if $a, b \in K$ then $ab - ba \in M$, whence $2(ab - ba) = 0$.

Also, $(2^{n-1} - 1)a \in M$ and $(2^{m-1} - 1)b \in M$ for suitable $n > 1$, $m > 1$ if $a, b \in K$. Moreover, $M$ is of characteristic 2, so by

Theorem 3.3.3 M is commutative. Hence $(2^{n-1} - 1)a$ commutes with $(2^{m-1} - 1)b$. Together with $2(ab - ba) = 0$, we deduce that $ab = ba$, finishing the proof of the theorem.

Similar arguments to these used above prove

THEOREM 3.3.7. Let R be any ring with involution in which $k^{n(k)} = k$, $n(k) > 1$, for every $k \in K_0$. Then any two elements in $K_0$ commute. Moreover, if J is the radical of R, then R/J is a subdirect product of fields and 2 × 2 matrix rings over fields.

## 4. Generalizations and a Theorem of Lee

In keeping with the notation used earlier, throughout this section polynomials written as $p_a(x)$ will indicate polynomials with integer coefficients which depend on a.

We now follow the general direction, laid out in the early theory of rings, in studying commutativity theorems. There, from the kind of results studied by us in the preceding sections one went to rings R where $x - x^2 p_x(x) \in Z$ for all $x \in R$. The theorem proved in that context was that R must be commutative [4].

In this section we shall study rings with involution in which $a - a^2 p_a(a) \in Z$ for various types of elements a — for instance for all $a \in S$ or $a \in T$, or for $a \in K$ or $a \in K_0$. For division rings we saw, in Theorems 3.2.4 and 3.2.5, that division rings so conditioned had to be very special, namely, at most 4-dimensional over their centers.

We shall see that for semi-simple rings a fairly definitive structure theorem can be given. We also obtain a fairly general condition under which all skew elements will commute.

We begin this material with

LEMMA 3.4.1. Let R be a *-prime; suppose that either

1. $t = t^2 p_t(t)$ for all $t \in T$,

or

2. $k = k^2 p_k(k)$ for all $k \in K_0$.

Then either R is commutative or is an order in a simple algebra 4-dimensional over its center. Thus R satisfies the standard identity $s_4(x_1, x_2, x_3, x_4)$. Furthermore, if R satisfies Condition 2, then any two elements in $K_0$ must commute.

Proof. If R is *-prime, but not prime, then there is an ideal $A \neq 0$ in R such that $AA^* = A^*A = 0$. If $a \in A$ then, if $a + a^* = (a + a^*)^2 p(a + a^*)$ or if $a - a^* = (a - a^*)^2 p(a - a^*)$ (according as the hypothesis imposed), we get $a = a^2 p(a)$. By the result quoted above, A is commutative. Hence $A + A^*$ is a commutative, non-zero *-ideal in R, which is *-prime. The outcome of this is that R must be commutative. Thus we may assume that R is prime.

If $K_0 = 0$, R is commutative. If $T = 0$, but $K_0 \neq 0$, then char $R \neq 2$ and $x = -x^*$ for all $x \in R$. This gives the contradiction $x^2 = 0$ for all $x \in R$, for $-(x^*)^2 = -(x^2)^* = (-x^*)^2 = x^{*2}$. So we may assume that $K_0 \neq 0$, $T \neq 0$. The proof of Theorem 3.2.3 shows that char $R = p \neq 0$. Also, our condition on T (or on $K_0$) shows that no non-zero element of T (or of $K_0$) is nilpotent.

We claim that $J(R) = 0$. Since $J(R)^* = J(R)$, if $a \in J(R)$ then $a^* \in J(R)$ and so $a \pm a^*$ are both in $J(R)$. In case of Condition 2, since $a - a^* = (a - a^*)^2 p(a - a^*)$ and $a - a^* \in J(R)$, we have $a = a^*$. Therefore $J(R)$ is commutative. If $J(R) \neq 0$ this implies that R is commutative.

If the condition is on $T$ (Condition 1), but not on $K_0$, then char $R \neq 2$, and $a + a^* = 0$ for all $a \in J(R)$. Thus we get $a^2 = 0$ for all $a \in J(R)$, leading to nilpotent ideals in $R$, if $J(R) \neq 0$. In short, $J(R) = 0$.

If $a \in T$ (or, $a \in K_0$) then, by our hypothesis, $a$ is algebraic over $GF(p)$, whence $a^m = a^n$ for some $0 < m < n$. Thus $a^{2m+1} = a^{2n+1}$ and so $(a^{2(n-m)+1} - a)^{2m+1} = 0$. However $a^{2(n-m)+1} = a^{(n-m)} a a^{(n-m)}$ is in $T$, or $K_0$, if $a$ is in $T$, or $K_0$. Thus $a^{2(n-m)+1} - a \in T$ (or, in $K_0$); since this element is nilpotent, $a^{2(n-m)+1} = a$ follows. By Theorems 3.3.2 (for Condition 1) or 3.3.7 (for Condition 2), $R$ satisfies $s_4(x_1, \ldots, x_4)$. By Posner's theorem, $R$ must be an order in a simple algebra which is 4-dimensional over its center.

Also, by Theorem 3.3.7 — in case of Condition 2 — any two elements of $K_0$ commute.

Since any semi-prime ring with involution is a subdirect product of $*$-prime rings, and since Conditions 1 and 2 are preserved under $*$-homomorphisms we immediately get from Lemma 3.4.1 that

THEOREM 3.4.1. Let $R$ be a semi-prime ring with involution. If either

1. $t = t^2 p_t(t)$ for every $t \in T$

or

2. $k = k^2 p_k(k)$ for every $k \in K_0$,

then $R$ is a subdirect product of commutative rings and orders in 4-dimensional simple algebras. Moreover, $R$ satisfies $s_4(x_1, x_2, x_3 . x_4)$. If the condition is on $K_0$, then any two elements in $K_0$ commute.

If in addition to our hypotheses above, we insist that $2R = R$ we can say a little more. For if $k \in K$ then $k = 2x$; hence $2x^* = k^* = -k = -2x$, whence $2(x+x^*) = 0$. But $x+x^* = 2r$ for some $r$, so that $0 = 2(x+x^*) = 4r$. Since $R$ is semi-prime, if $4r = 0$ then $2r = 0$. Therefore, $x = -x^*$ and so $k = x - x^*$, which is to say, $K = K_0$. Thus in the presence of Condition 2, we get that any two elements of $K$ (not just of $K_0$) commute.

THEOREM 3.4.2. Let $R$ be any ring with involution in which $k = k^2 p_k(k)$ for $k \in K_0$. Then any two elements of $K_0$ commute. If $2R = R$, then any two elements of $K$ commute.

Proof. Let $N$ be the maximal nil ideal of $R$ and suppose $k_1, k_2 \in K_0$. In $\overline{R} = R/N$, $\overline{k}_1$ and $\overline{k}_2$ are skew-traces; by Theorem 3.4.1, $\overline{k}_1\overline{k}_2 = \overline{k}_2\overline{k}_1$, whence $k_1k_2 - k_2k_1 \in N$, so is nilpotent. Because $k_1k_2 - k_2k_1 \in K_0$ and, by the hypothesis, $K_0$ has no non-zero nilpotent elements, we have $k_1k_2 = k_2k_1$.

If $2R = R$ and $a, b \in K$ then, in $\overline{R} = R/N$, $\overline{a}\,\overline{b} = \overline{b}\,\overline{a}$ by the remarks preceding the theorem. Hence $ab - ba \in N$; but $ab - ba \in K_0$. The net result of these is that $ab = ba$.

We now wish to weaken the conditions of the two preceding theorems. Our first step in this direction is

THEOREM 3.4.3. Let $R$ be a primitive ring with involution. Suppose that either

1. $s - s^2 p_s(s) \in Z = Z(R)$ for all $s \in S$

or

2. $k - k^2 p_k(k) \in Z$ for all $k \in K$.

Then R is a simple algebra with $\dim_Z R \leq 4$. In case of Condition 1, $S \subset Z$; in Condition 2, any two elements of K commute. If char R = 2 then R must be a field.

Proof. If R is a division ring then the result is correct by applying Theorem 3.2.4 or Theorem 3.2.5. So we may assume that R is not a division ring.

The hypothesis implies that no non-zero element of S (or of K) is nilpotent — since Z has no nilpotent elements; thus, by Theorem 2.2.1 (or Theorem 2.2.4) either $xx^* \neq 0$ if $x \neq 0$, or R is an order in $F_2$. In the latter case, since R is primitive and satisfies the identities of $F_2$, we get R is the ring of $2 \times 2$ matrices over a field. Then it is trivial to verify the rest of the theorem. So we may assume that $xx^* \neq 0$ if $x \neq 0$.

Suppose that R has a symmetric idempotent $e \neq 0, 1$. If $s - s^2 p_s(s) \in Z$ for all $s \in S$, then if $\lambda^* = \lambda \in Z$, $\lambda e \in S$ hence $\lambda e - (\lambda e)^2 p_{\lambda e}(\lambda e) \in Z$; that is, $(\lambda - \lambda^2 p_{\lambda e}(\lambda))e \in Z$. But no element of Z is a sero divisor; therefore $\lambda - \lambda^2 p_{\lambda e}(\lambda) = 0$, whence $\lambda$ is algebraic over the prime field. Using the argument of Theorem 3.2.3 now, we get char $R = p \neq 0$. Hence $\lambda$ is algebraic over $GF(p)$; from this we get that Z is algebraic over $GF(p)$, and so S is algebraic over $GF(p)$. Thus, if $s \in S$, $s^m = s^n$ with $1 \leq m < n$ which gives us $(s^{n-m+1} - s)^m = 0$. Because $s^{n-m+1} - s$ is nilpotent and symmetric, we get $s^{n-m+1} = s$. Thus by Theorem 3.3.1 we obtain the present theorem.

On the other hand, if $k - k^2 p_k(k) \in Z$ for $k \in K$, since for char R = 2 we have the result by the above, we may assume that char $R \neq 2$. If $\mu^* = -\mu \neq 0 \in Z$, then $\mu e \in K$ whence

$\mu e - \mu^2 e p_{\mu e}(\mu e) \in Z$, that is, $(\mu - \mu^2 p(\mu))e \in Z$. As above, this gives us $\mu = \mu^2 p(\mu)$, and so $\mu$ is algebraic over the prime field. If $\lambda^* = \lambda \in Z$ then $(\lambda\mu)^* = -\lambda\mu$, so $\lambda\mu$ is algebraic over the prime field, hence $\lambda$ is. Thus $Z$ is algebraic over the prime field. Also from the form of the polynomial satisfied by $\mu$, $2\mu$ we get, as in Theorem 3.2.3, that char $R = p \neq 0$. As above we conclude that $k^m = k$ for some $m = m(k) > 1$, hence by Theorem 3.3.7 we would be done. So to finish here we must know that there is a $\mu \in Z$, $\mu^* = -\mu \neq 0$. But $b = k - k^2 p(k) \in Z$ hence $\mu = b - b^* = 2k - k^2(p(k) - p(-k)) \in Z$ and is skew. So if $\mu = 0$ for all $k \in K$, we get $2k = k^2(p(k) - p(-k))$; because $2K = K$, we get $a = a^2 q_a(a)$ for all $a \in K$. By Lemma 3.4.1 the result follows.

So, to finish the proof, we merely must show that $R$ has a symmetric idempotent $e \neq 0, 1$. Because $R$ is primitive and not a division ring, it contains a maximal regular right ideal $\rho \neq 0$ which contains no non-zero two-sided ideal of $R$; hence $\rho \cap Z = 0$. If $x \neq 0 \in \rho$ then $s = xx^* \neq 0$ and is in $\rho$. Thus $sRs \subset \rho$, whence $sRs \cap Z = 0$. However $sRs$ is primitive — this because $xx^* \neq 0$ for $x \neq 0$, and Lemma 2.3.2; moreover if $u \in S \cap sRs$ (or $v \in K \cap sRs$) then $u - u^2 p_u(u) \in Z$ forces $u = u^2 p_u(u)$ (or $v = v^2 p_v(v)$). By Lemma 3.4.1, $sRs$ satisfies a polynomial identity; being primitive, $sRs$ must be a finite-dimensional simple algebra. Thus $sRs$ has a unit element $e \neq 0$; $e$ is an idempotent, $e^* = e$ and $e \neq 1$, for $s \in \rho$ is not invertible. This finishes the proof.

This last theorem is of a rather definitive character — at least for primitive rings; yet it suffers from one great defect. Suppose we wanted to investigate the structure of semi-simple rings $R$ satisfying the conditions of the theorem on $S$ or on $K$. We would then look at the primitive images of this ring. However, the conditions on $S$ or on $K$ might not carry over to the primitive image; this could happen only if the primitive image happens to be of characteristic 2, in which case its symmetric elements need not be images of symmetric elements in R. For this reason we want to study this type of condition which is preserved under *-homomorphisms. Because traces and skew-traces in a *-homomorphic image of a ring $R$ come from traces and skew-traces in R, we now study the situation where the conditions of Theorem 3.4.3 are imposed only on $T$ or on $K_0$.

Because of the remarks above we clearly need but concentrate on the case of characteristic 2; here, of course, $K_0 = T$.

LEMMA 3.4.2. Let $E \supsetneq F$ be fields of characteristic 2, and suppose that for each $a \in E$, $a + a^2 p_a(a) \in F$. Then $E$ is algebraic over $GF(2)$.

<u>Proof.</u> By a theorem of Herstein [6] concerning three fields, either $E$ is algebraic over $GF(2)$ or every element in $E$ is purely inseparable over $F$. Suppose the latter; then there is an $a \in E$, $a \notin F$, such that $a^2 \in F$. However, $a + a^2 p_a(a) = a + n_2 a^2 + n_3 a^3 + \ldots + n_k a^k$ is in $F$; together with $a^{2m} \in F$ we have $a(1 + n_3 a^2 + n_5 a^4 + \ldots) \in F$. Since $a \notin F$ and $1 + n_3 a^2 + n_5 a^4 + \ldots \in F$ we must have $1 + n_3 a^2 + n_5 a^4 + \ldots = 0$, that is, $a$ is algebraic over $GF(2)$. Hence $a^{2^t} = a$ for some $t$; because $a^2 \in F$ we thus get $a = a^{2^t} \in F$, a contradiction.

This allows us to prove

LEMMA 3.4.3. Let $D$ be a division ring with involution, of characteristic 2. Suppose that $t + t^2 p_t(t) \in Z$ for all $t \in T$. Then $\dim_Z D \leq 4$.

Proof. If every $t \in T$ is algebraic over $GF(2)$ then we are done by Theorem 3.1.2. So we may assume, for some $t \in T$, that
$\lambda = t + t^2 p_t(t) \neq 0 \in Z$. We can rewrite
$\lambda = t + t^2 p_t(t) = t(1 + ta(t))^2 + t^2 b(t^2)$ where the polynomials $a$ and $b$ have integer coefficients. Since $t$ is not algebraic over $GF(2)$, $1 + ta(t) \neq 0$; hence, if $u = (1 + ta(t))^{-1}$ then $u^* = u$ and $t = u^{-1}(\lambda + t^2 b(t^2) u^{-1}$, whence $t^2 = u^{-1}(\lambda t + t^3 b(t^2))u^{-1}$. But $\lambda t, t^3$ and $t^3 b(t^2)$ are in $T$, thus $t^2 \in T$ results. Therefore we get that $\lambda \neq 0$ is in $Z \cap T$.

Because $\lambda = a + a^*$ for some $a \in R$, $1 = \lambda^{-1} a + \lambda^{-1} a^* = \lambda^{-1} a + (\lambda^{-1} a)^*$, hence $1 \in T$, and so $Z \cap S \subset T$. But then $x1x^* = xx^* \in xTx^* \subset T$; in particular, all powers of any trace are in $T$.

If $Z^+ = Z \cap S$ and $t_0 \in T$, $t_0 \notin Z$ then $E = Z^+(t_0) \supsetneqq F = Z^+$ and every element $a$ in $E$ — being a polynomial in $t_0$ over $Z^+$ — satisfies a polynomial of the form $a + a^2 p_a(a) \in F$. By the previous lemma, $E$ is algebraic over $GF(2)$. From this $Z$ is algebraic over $GF(2)$, whence $T$ must be. By Theorem 3.1.2, we obtain the result.

From the division rings we quickly pass to the case of primitive rings.

LEMMA 3.4.4. Let $R$ be a primitive ring with involution, of characteristic 2. Suppose that $t + t^2 p_t(t) \in Z$ for all $t \in T$; then $R$ is simple and $\dim_Z R \leq 4$.

Proof. If $R$ is a division ring the result is merely that of Lemma 3.4.3; so we may suppose that $R$ is not a division ring. Hence $R$ has a maximal regular right ideal $\rho \neq 0$ which contains no non-zero ideal of $R$; in particular, $\rho \cap Z = 0$.

The hypothesis implies that no non-zero element of $T$ is nilpotent; by Theorem 2.2.4 we have that either $R$ is an order in $F_2$, or $xx^* \neq 0$ if $x \neq 0$ in $R$. The first possibility implies that $R$ is the ring of $2 \times 2$ matrices over a field. So we may assume that $xx^* \neq 0$ if $x \neq 0$.

If $x \neq 0 \in \rho$ then $s = xx^* \neq 0$ is in $\rho$ hence $sTs \subset \rho$. If $T \cap Z \neq 0$ then $t = \lambda s^2 \in \rho \cap T$ for $\lambda \neq 0$ in $T \cap Z$, hence $t + t^2 p_t(t) \in Z \cap \rho \neq 0$. This gives $t = t^2 p_t(t)$ and so $e = tp_t(t) \neq 0$ is a symmetric idempotent, because $e \in \rho$, $e \neq 1$. The proof of Theorem 3.4.3 now carries over to prove the result.

Thus, to finish, it is sufficient to prove that $Z \cap T \neq 0$. Since there are no nilpotent elements in $T$, by Corollary 2 to Theorem 3.1.2, there is a $t \in T$ such that $t$ is not algebraic over $GF(2)$. But $0 \neq \lambda = t + t^2 p(t) = tq_1(t^2) + q_2(t^2)t^2 \in Z$. If $t_1 = tq_1(t^2)$ then $t_1 \in T$ since it is a sum of odd powers of $t \in T$. Also, $t_1^2 = t_1 t_1 = tq_1(t^2)(\lambda + q_2(t^2)t^2)$ is a sum of $\lambda t q_1(t^2)$, which is in $T$, and $tq_1(t^2)q_2(t^2)t^2$, which is a sum of odd powers of $t$ so is also in $T$. Hence $t_1^2 \in T$. Therefore, all powers of $t_1$ are in $T$. But because $t_1 \in T$, $\mu = t_1 + t_1^2 q(t_1) \in Z$; by the above, $\mu \in T$. So, if $\mu \neq 0$, then $Z \cap T \neq 0$. If $\mu = 0$ then $t_1$ is algebraic over $GF(2)$, hence $tq_1(t^2)$ is algebraic over $t_1$, whence $t$ is — a contradiction. The lemma is now proved.

This allows us to push Theorem 3.4.3 to a sharper form.

THEOREM 3.4.4. Let $R$ be a primitive ring with involution in which either

1. $t - t^2 p_t(t) \in Z$ for all $t \in T$,

or

2. $k - k^2 p_k(k) \in Z$ for all $k \in K_0$.

Then $R$ is simple and $\dim_Z R \leq 4$. Moreover, in case of Condition 2, any two elements of $K_0$ commute.

Proof. If $\operatorname{char} R \neq 2$, the result is implied by Theorem 3.4.3. If $\operatorname{char} R = 2$ this is merely Lemma 3.4.4.

From here it is standard to get

THEOREM 3.4.5. Let $R$ be a semi-simple ring with involution in which either

1. $t - t^2 p_t(t) \in Z$ for all $t \in T$,

or

2. $k - k^2 p_k(k) \in Z$ for all $k \in K_0$.

Then $R$ is a subdirect product of simple algebras which are at most 4-dimensional over their centers. Moreover, in case of Condition 2, any two elements of $K_0$ commute.

If $R$ is semi-prime and $2R = R$ then we saw that $K_0 = K$. The condition $2R = R$ and $K_0$ are preserved under *-homomorphism. Thus we have

THEOREM 3.4.6. Let $R$ be a semi-simple ring with involution in which $k - k^2 p_k(k) \in Z$ for all $k \in K_0$. If $2R = R$ then any two elements of $K$ commute.

A few remarks before closing. The preceding results for S are special cases of more general results in a paper [2] by Chacron, Herstein, and Montgomery. They work out the structure of rings in which, for a <u>fixed</u> k, $s^k - s^{k+1}p_s(s) \in Z$ for all $s \in S$. All the preceding results pertaining to $K_0$ and K are due to P.H. Lee [12].

Lee and Herstein, in a joint paper [7] have the result of Theorem 3.4.6 without any assumption of semi-simplicity. Their result is:

Let R be a ring with involution such that $2R = R$. If, for every $k \in K_0$, $k - k^2p_k(k) \in Z$ then any two elements of K must commute.

Bibliography

1. M. Chacron. A generalization of a theorem of Kaplansky and rings with involution. Mich. Math. Jour. 20 (1973): 45-54.

2. M. Chacron, I. N. Herstein and Susan Montgomery. Structure of a certain class of rings with involution. (To appear)

3. I. N. Herstein. Non-Commutative Rings. Carus Monograph 15, 1968.

4. I. N. Herstein. Structure of a certain class of rings. Amer. Jour. Math. 75 (1953): 866-871

5. I. N. Herstein. Rings with periodic symmetric on skew elements. Jour. Algebra 30 (1974): 144-154.

6. I. N. Herstein. A theorem concerning three fields. Canadian Jour. 7 (1955): 202-203.

7. I. N. Herstein and P. H. Lee. Commuting skew elements in rings with involution. (To appear)

8. I. N. Herstein and S. Montgomery. A note on division rings with involution. Mich. Math. Jour. 18 (1971): 75-79.

9. N. Jacobson. Structure of Rings. AMS Colloq. Publications XXXVII ($2^{nd}$ ed.) 1964.

10. N. Jacobson. Lie Algebras. Interscience 1962.

11. I. Kaplansky. A theorem on division rings. Canadian Jour. Math. 3 (1951): 290-292.

12. P. H. Lee. On primitive rings with involution. Jour. Algebra 32 (1974): 611-622.

13. S. Montgomery. A generalization of a theorem of Jacobson. Proc. AMS 28 (1971): 366-370.

14. S. Montgomery. A generalization of a theorem of Jacobson II. Pacific Jour. Math. 44 (1973): 233-240.

15. M. Nagata, T. Nakayama and T. Tuzuku. On an existence lemma in valuation theory. Nagoya Math. Jour. 6 (1953): 59-61.

16. J.M. Osborn. Varieties of algebras. Advances in Math. 8 (1972): 163-369.

# 4

# MAPPING THEOREMS

In this chapter we shall study some mappings which preserve certain combinations of elements in a ring with involution, elements which reflect that the ring in question has an involution. Our aim, in general, is to show that these mappings must be very well-behaved, for example, must be homomorphisms, derivations, and the like.

## 1. Some Results of Lynne Small

We shall develop here some of the theorems obtained by Lynne Small in [7]. She used these results to exhibit the non-isomorphism of certain irreducible Jordan modules of R and S, where R is a simple ring with involution and S the set of symmetric elements of R. We shall only touch on this module aspect lightly, referring the reader to Small's paper for these. Our emphasis will be more in the direction of the nature of the maps and the mapping theorems. Our first result in this direction is

THEOREM 4.1.1. Let R be a simple ring with involution, of characteristic not 2. Let $\varphi : R \to A$ be an additive mapping such that $\varphi(xx^*) = \varphi(x)\varphi(x^*)$ for every $x \in R$. If $\dim_Z R > 4$, and if $\overline{\varphi(R)}$, the subring of A generated by $\varphi(R)$, is semi-prime, then $\varphi$ is a homomorphism of R into A.

Proof. First of all note that if $a \in S$ or $a \in K$ then $\varphi(a^2) = \varphi(a)^2$. Linearize $\varphi(xx^*) = \varphi(x)\varphi(x^*)$ by replacing $x$ by $x + y^*$; we get, in this way,

(1) $$\varphi(xy + y^*x^*) = \varphi(x)\varphi(y) + \varphi(y^*)\varphi(x^*), \quad \text{all } x, y \in R.$$

As a special case of (1), using $x = s \in S$ and $y = k \in K$, we have

(2) $$\varphi(sk - ks) = \varphi(s)\varphi(k) - \varphi(k)\varphi(s).$$

We compute $\varphi((xy + y^*x^*)x + x^*(xy + y^*x^*))$ in two ways; first by using (1) twice on $z = (xy + y^*x^*)x + x^*(xy + y^*x^*)$, and second, by noting that $z = xyx + x^*y^*x^* + (xy)^*x + x^*(xy)$. What comes out in comparing these two computations is

(3) $$\varphi(xyx + x^*y^*x^*) = \varphi(x)\varphi(y)\varphi(x) + \varphi(x^*)\varphi(y^*)\varphi(x^*), \text{ for all } x, y \in R.$$

In particular, if $x \in K$ and $y \in S$ in (3), we get that $2\varphi(ksk) = 2\varphi(k)\varphi(s)\varphi(k)$ for $k \in K$, $s \in S$, and since $2S = S$, $2K = K$ in $R$, we get from this equality above that

(4) $$\varphi(ksk) = \varphi(k)\varphi(s)\varphi(k) \quad \text{for } k \in K, s \in S.$$

Finally, using the special values $x = s + k$, $y = k_1$, where $s \in S$, $k, k_1 \in K$, in (3) gives us

(5) $$\varphi(skk_1 + k_1ks) = \varphi(s)\varphi(k)\varphi(k_1) + \varphi(k_1)\varphi(k)\varphi(s).$$

Define, for $x, y \in R$, $x^y = \varphi(xy) - \varphi(x)\varphi(y)$. Our objective is to prove that $x^y = 0$ for all $x, y \in R$, that is, that $\varphi$ is a homomorphism. Note that (2) tells us that $k^s = s^k$ for $k \in K$, $s \in S$.

We compute. If $s \in S$, $k, k_1 \in K$ then, by (1), $\varphi((sk)k_1 + k_1(ks)) = \varphi(sk)\varphi(k_1) + \varphi(k_1)\varphi(ks)$; subtracting this last result from that of (5), we wind up with $s^k\varphi(k_1) + \varphi(k_1)k^s = 0$, and since $k^s = s^k$, we have that $k^s\varphi(k_1) + \varphi(k_1)k^s = 0$. Hence $k^s\varphi(k_1)^2 = \varphi(k_1)^2k^s$ and so $k^s\varphi(k_1^2) = \varphi(k_1^2)k^s$. By Theorem 2.1.11, since $\dim_Z R > 4$, if $t \in S$

then $t = \sum k_i^2 - \sum k_j'^2$ where $k_i, k_j' \in K$. Thus, by the above $k^s\varphi(t) = \varphi(t)k^s$ for all $t \in S$. This, together with $k^s\varphi(k_1) = -\varphi(k_1)k^s$ for $k_1 \in K$, gives us

(6) $$k^s\varphi(x) = \varphi(x^*)k^s \quad \text{for} \quad x \in R,\ k \in K,\ s \in S.$$

We shall now show that $(k^s)^2 = 0$. This is a consequence of a long, easy computation that we now perform.

$$\begin{aligned}
(k^s)^2 &= k^s s^k = k^s\varphi(sk) - k^s\varphi(s)\varphi(k) \\
&= -\varphi(ks)k^s - k^s\varphi(s)\varphi(k) \qquad \text{(using } x = sk \text{ in ( 6 ))} \\
&= -\varphi(ks)s^k - k^s\varphi(s)\varphi(k) \\
&= \varphi(ks)\varphi(s)\varphi(k) - \varphi(sk)) + (\varphi(k)\varphi(s) - \varphi(ks))\varphi(s)\varphi(k) \\
&= -\varphi(ks)\varphi(sk) + \varphi(k)\varphi(s)^2\varphi(k) \\
&= \varphi(ks)\varphi((ks)^*) + \varphi(ks^2k) \quad , \quad \text{by (4)} \\
&= \varphi(ks(ks)^* + ks^2k) \\
&= \varphi(-ks^2k + ks^2k) = 0 .
\end{aligned}$$

Now that we know that $(k^s)^2 = 0$ and that $k^s\varphi(x) = \varphi(x\ )k^s$, it is clear that $k^s\overline{\varphi(R)}$ is a nilpotent right ideal of $\overline{\varphi(R)}$. By hypothesis we must have that $k^s = 0$, which is to say, $\varphi(ks) = \varphi(k)\varphi(s)$, for all $k \in K$, $s \in S$. By (2) we then also have that $\varphi(sk) = \varphi(s)\varphi(k)$.

Since $sk - ks \in S$ we have

$$\begin{aligned}
\varphi((sk - ks)k) &= \varphi(sk - ks)\varphi(k) \\
&= (\varphi(s)\varphi(k) - \varphi(k)\varphi(s))\varphi(k) \\
&= \varphi(s)\varphi(k)^2 - \varphi(k)\varphi(s)\varphi(k) \\
&= \varphi(s)\varphi(k)^2 - \varphi(ksk) \qquad \text{by (4)} .
\end{aligned}$$

Therefore we see that $\varphi(sk^2) = \varphi(s)\varphi(k^2)$. Using Theorem 2.1.11 again, we obtain from this last relation that $\varphi(st) = \varphi(s)\varphi(t)$ for all

$s, t \in S$. Since $\varphi(sk) = \varphi(s)\varphi(k)$ for $k \in K$, we now have $\varphi(sk) = \varphi(s)\varphi(x)$ for all $x \in R$, $s \in S$. Thus we obtain, by iteration, $\varphi(s_1 s_2 \ldots s_n x) = \varphi(s_1)\varphi(s_2)\ldots\varphi(s_n)\varphi(x) = \varphi(s_1 s_2 \ldots s_n)\varphi(x)$ for all $s_1, \ldots, s_n \in S$, and $x \in R$. Since $\dim_Z R > 4$, $S$ generates $R$, and so we see that $\varphi(yx) = \varphi(y)\varphi(x)$ for all $x, y \in R$. In other words, $\varphi$ is a homomorphism of $R$ into $A$.

Since $R$ is simple, if $\varphi \neq 0$ then $\varphi$ must therefore be one-to-one.

Note that in the proof we did not use the full force of the hypothesis that $\overline{\varphi(R)}$ is semi-prime. All we needed was that $k^s\overline{\varphi(R)}$ was not nilpotent. This remark is relevant for the next theorem which gives an an analog of Theorem 4.1.1 for derivations.

THEOREM 4.1.2. Let $R$ be a simple ring with involution, of characteristic not 2, such that $\dim_Z R > 4$. Let $\delta: R \to R$ be such that $\delta(xx^*) = x\delta(x^*) + \delta(x)x^*$ for all $x \in R$. Then $\delta$ is a derivation of $R$.

Proof. We use an old trick of going to $2 \times 2$ matrices. Let $A = \left\{ \begin{pmatrix} a & c \\ 0 & b \end{pmatrix} \mid a, b, c \in R \right\}$; $A$ is a ring. Define $\varphi: R \to A$ by

$$\varphi(a) = \begin{pmatrix} a & \delta(a) \\ 0 & a \end{pmatrix}$$ for $a \in R$. Using the property

$\delta(aa^*) = a\delta(a^*) + \delta(a)a^*$ we see that $\varphi(aa^*) = \varphi(a)\varphi(a^*)$ for every $a \in R$. In the proof of Theorem 4.1.1 we saw that if $k \in K$, $s \in S$ then $k^s\varphi(x) = \varphi(x^*)k^s$ for all $x \in R$, where $k^s = \varphi(ks) - \varphi(k)\varphi(s)$. We write this equality out explicitly; it gives $(\delta(ks) - \delta(k)s - k\delta(s))x = x^*(\delta(ks) - \delta(k)s - k\delta(s)$ for all $k \in K$, $s \in S$, $x \in R$. If $b = \delta(ks) - \delta(k)s - k\delta(s)$, we have that $bx = x^*b$ for all $x \in R$. Hence $bxy = x^*by = x^*y^*b = (yx)^*b = byx$, that is, $b[R, R] = 0$. The simplicity of R forces $b = 0$. Thus $\delta(ks) = k\delta(s) + \delta(k)s$ for $k \in K$, $s \in S$. But then $k^s = 0$ follows. From this point on — once we know $k^s = 0$ — the proof of Theorem 4.1.1 goes through as before to show that $\varphi(xy) = \varphi(x)\varphi(y)$ for all $x, y \in R$. This says that

$$\begin{pmatrix} xy & \delta(xy) \\ 0 & xy \end{pmatrix} = \begin{pmatrix} x & \delta(x) \\ 0 & x \end{pmatrix}\begin{pmatrix} y & \delta(y) \\ 0 & y \end{pmatrix}$$

hence $\delta(xy) = x\delta(y) + \delta(x)y$. In other words, $\delta$ is a derivation of R. This proves the theorem.

We now prove another result in the spirit of Theorem 4.1.2. This is

THEOREM 4.1.3. Let R be a simple ring with involution $*$ of characteristic not 2. Suppose that $\delta : R \to R$ is an additive mapping such that $\delta(xx^*) = x\delta(x)^* + \delta(x)x^*$ for all $x \in R$. If $\dim_Z R > 4$, $\delta$ must be a derivation of R.

Proof. In light of Theorem 4.1.2 we must but show that $\delta(x^*) = \delta(x)$ for every $x \in R$.

From its form it is clear that $\delta(xx^*)$ is symmetric, and in particular, $\delta(s^2)$ is symmetric if $s \in S$. But, since char $R \neq 2$, the additive group generated by all $s^2$, $s \in S$, is merely S. Hence we get that $\delta(t)$ is symmetric for $t \in S$. In particular, $\delta(s^2) = s\delta(s)^* + \delta(s)s$ $= s\delta(s) + \delta(s)s$, for every $s \in S$.

In the relation $\delta(xx^*) = x\delta(x)^* + \delta(x)x^*$ let $x = s + k$ where $s \in S$, $k \in K$. Using that $\delta(t)^* = \delta(t)$ for $t \in S$ and $-\delta(k^2) = \delta(kk^*) = k\delta(k)^* - \delta(k)k$, we get from the above linearization

(1) $\delta(ks - sk) = k\delta(s) - \delta(s)k + s\delta(k)^* + \delta(k)s$ .

Computing $\delta(ks^2 - s^2k)$ according to (1), and computing it from the fact that $ks^2 - s^2k = (ks - sk)s + s(ks - sk)$, and knowing the behavior of $\delta$ on the Jordan products of symmetric elements, we get

(2) $$s(\delta(k)^* + \delta(k))s = 0 \quad \text{for } s \in S, k \in K.$$

Let $W = \{w \in S \mid sws = 0 \text{ all } s \in S\}$. We claim that $W$ is a Jordan ideal of $S$. For if $w \in W$ and $s, t \in S$ then from $(s+t)w(s+t) = 0$ we see that $swt + tws = 0$. Replacing $t$ by $su + us$ where $u \in S$ leads, on applying $sws = 0$, to $s(wu + uw)s = 0$; in other words, $W$ is a Jordan ideal of $S$. Thus, by Theorem 2.1.12 $W = 0$ or $W = S$.

We rule out the possibility $W = S$. In that case, since $sSs = 0$ for every $s \in S$, if $x = t+k \in R$, $t \in S$, $k \in K$ then $sxs = sks$ and $sxsxs = sksks = 0$ since $ksk \in S$. Thus $sR$ is nil of index 3, which is not possible in a simple ring. Thus $W = 0$. But $\delta(k)^* + \delta(k) \in W$ for $k \in K$, hence $\delta(k)^* = -\delta(k) = \delta(k^*)$. Since $x = s+k$, $s \in S$, $k \in K$, for any $x \in R$ then $\delta(x)^* = (\delta(s) + \delta(k))^* = \delta(s)^* + \delta(k)^* = \delta(s) - \delta(k) = \delta(s - k) = \delta(x^*)$. The result now follows from Theorem 4.1.2.

One might wonder why we do not prove a multiplicative analog for Theorem 4.1.3, namely for the mapping $\varphi$ such that $\varphi(xx^*) = \varphi(x)\varphi(x)^*$. It is easy enough — even on $3 \times 3$ matrices — to give examples of additive maps which are not homomorphisms where $0 = \varphi(xx^*) = \varphi(x)\varphi(x)^*$ for all $x \in R$.

We prove one more theorem in this vein. This result can be used to exhibit certain non-isomorphic irreducible Jordan modules for a simple ring with involution.

THEOREM 4.1.4. Let $R$ be a simple ring with involution, of characteristic not 2. Let $\varphi: S \to R$ be an additive mapping such that $\varphi(st + ts) = s\varphi(t) + \varphi(t)s$ for all $s, t \in S$. If $\dim_Z R > 4$ then $\varphi(s) = \lambda s$ where $\lambda$ is in the centroid of $R$.

Proof. The condition on $\varphi$ is symmetric, so we see that $s\varphi(t) + \varphi(t)s = t\varphi(s) + \varphi(s)t$. In this relation, replace $t$ by $st + ts$; we get $s\varphi((st + ts)) + \varphi(st + ts)s = (st + ts)\varphi(s) + \varphi(s)(st + ts)$. Using $\varphi(st + ts) = \varphi(s)t + t\varphi(s)$ and comparing the computation, we get that $t[s, \varphi(s)] = [s, \varphi(s)]t$ for all $s, t \in S$. Since $\dim_Z R > 4$, $S$ generates $R$, hence $[s, \varphi(s)] \in Z$.

If $Z = 0$ this gives us that $s\varphi(s) = \varphi(s)s$ for all $s \in S$. Since $2\varphi(s^2) = \varphi(ss + ss) = s\varphi(s) + \varphi(s)s = 2s\varphi(s)$, and since char $R \neq 2$, $\varphi(s^2) = s\varphi(s)$ for $s \in S$. Linearizing this gives $s\varphi(t) + t\varphi(s) = \varphi(st + ts) = s\varphi(t) + \varphi(t)s$, hence $t\varphi(s) = \varphi(t)s$ for all $s, t \in S$. If $u \in S$ then $\varphi(t)su = t\varphi(s)u = ts\varphi(u)$. Continuing this way, we get $\varphi(t)\bar{s}u = t\bar{s}\varphi(u)$ for any $\bar{s}$ in the subring generated by $S$. Because $\dim_Z R > 4$, the subring generated by $S$ is $R$; therefore $\varphi(t)xu = tx\varphi(u)$ for all $x \in R$. Using $u = t$, we see that $tx\varphi(t) = \varphi(t)xt$ for all $x \in R$. By Lemma 1.3.2 $\varphi(t) = \lambda_t t$ where $\lambda_t$ is in the extended centroid, $C$, of $R$; but, because $R$ is simple, $C$ is merely the centroid of $R$. To finish, we merely must show that we can pick $\lambda_t$ independent of $t$.

If $t \neq 0 \in S$, let $\lambda_t$ be as above and let $U = \{s \in S \mid \varphi(s) = \lambda_t s\}$. From the properties of $\varphi$ it is immediate that $U$ is a Jordan ideal of $S$, and $U \neq 0$ since $t \neq 0$ is in $U$. Therefore, by Theorem 2.1.12, $U = S$. Thus $\varphi(s) = \lambda s$ for all $s \in S$, where $\lambda = \lambda_t$ is in the centroid of $R$.

On the other hand, if $Z \neq 0$ then $1 \in R$. But $1 \in S$, hence $[s+1, \varphi(s+1)] \in Z$. This gives $\lambda = [s, \varphi(1)] \in Z$. Now $2\lambda s = s[s, \varphi(1)] + [s, \varphi(1)]s = [s^2, \varphi(1)] \in Z$. If $s \notin Z$ this forces $\lambda = 0$; if $s \in Z$ then certainly $\lambda = 0$. Therefore $\varphi(1)$ centralizes S and so must be in Z. But $2\varphi(s) = \varphi(1s + s1) = s\varphi(1) + \varphi(1)s = 2\varphi(1)s$, whence $\varphi(s) = \varphi(1)s$ where $\varphi(1) \in Z$. The theorem is not proved.

If $s \in S$, $t \in S$ then $st + ts \in S$; if $s \in S$, $k \in K$, then $sk+ks \in K$. So S and K are Jordan S-modules. When are they isomorphic as Jordan S-modules ?

THEOREM 4.1.5. Let R be simple with involution, of characteristic not 2, such that $\dim_Z R > 4$. Then S and K are isomorphic as Jordan S-modules if and only if the involution is of the second kind (on the centroid of R).

Proof. If the involution is of the second kind on the centroid of R, let $\lambda \neq 0 \in \mathfrak{Z}(R)$ such that $\lambda^* = -\lambda$. Then $\lambda S = K$. Define $\varphi: S \to K$ by $\varphi(s) = \lambda s$; it is trivial that $\varphi$ makes S and K isomorphic as Jordan S-modules.

On the other hand, suppose that S and K are isomorphic as Jordan S-modules. Then there is a mapping $\varphi: S \to K$ such that $\varphi(st + ts) = s\varphi(t) + \varphi(t)s$ for all $s, t \in S$. By Theorem 4.1.4, $\varphi(s) = \lambda s$ for all $s \in S$, where $\lambda \neq 0 \in \mathfrak{Z}(R)$. Since $K = \lambda S$, we must have $\lambda^* = -\lambda$, hence the involution is of the second kind on the centroid of R.

We know by Theorem 2.1.12 that S is a simple Jordan ring, and so is simple as a Jordan S-module, whenever R is simple of characteristic not 2. One can ask whether something similar is true for K as a Jordan S-module. The answer, in general, is yes [1].

We need a subsidiary result, which is really implicitly contained in the results of Section 1 of Chapter 2. However we prove it formally here.

LEMMA 4.1.1. Let R be a simple ring with involution, of characteristic not 2, and suppose that $\dim_Z R > 4$. Then $[S, S] \supset [K, K]$.

Proof. If $U = S + [S, S]$ then, clearly, $[U, S] \subset U$ and $[U, K] \subset U$. Since $R = S + K$, $[U, R] \subset U$, that is to say, U is a Lie ideal of R. Hence, by Theorem 2.1.3, either $U \subset Z$ of $U \supset [R, R]$. However, if $U \subset Z$ then we have $S \subset Z$, and this forces $\dim_Z R \leq 4$ by Theorem 2.1.5. Therefore $U = S + [S, S] \supset [R, R]$ hence $U \supset [K, K]$. But the skew part of U is $[S, S]$, hence $[S, S] \supset [K, K]$.

THEOREM 4.1.6. Let R be a simple ring, of characteristic not 2, with $\dim_Z R > 4$. If $U \subset K$ is an additive subgroup such that $U \circ S \subset U$ then $U = 0$ or $U = K$; that is, K is simple as a Jordan S-module.

Proof. If $u \in U$, $s \in S$ then, from $(us + su)s + s(su + us) \in U$, $s^2 \in S$, we get $2sus \in U$, hence $sus \in U$. Linearizing on s gives $sut + tus \in U$ for all $s, t \in S$, $u \in U$. However, $(su + us)t + t(su + us) \in U$, whence $ust + tsu \in U$. Interchanging s and t and subtracting we obtain $u(st - ts) - (st - ts)u \in U$, that is, $[U, [S, S]] \subset U$. Since $\dim_Z R > 4$, $[S, S] \supset [K, K]$ by Lemma 4.1.1, hence $[U, [K, K]] \subset U$. Also, $[[U, S], S] \subset U$ follows, since $(us - su)t - t(us - su) = ust + tsu - (sut + tus) \in U$ by the above.

Thus, if $V = U + [U, S]$ then $[V, S] \subset V$ and $[V, [K, K]] \subset V$, hence since $R = S + K$, $[V, [R, R]] \subset V$. By Theorem 2.1.4, either $V \subset Z$ or $V \supset [R, R]$. If $V \subset Z$ then $U \subset Z$, so if $U \neq 0$, the involution is of the second kind, and so $K$ is irreducible as an S-module. Suppose then that $V \supset [R, R]$, whence $U + [U, S] \supset S + [K, K]$. We get from this that $U \supset [K, K]$, and so $U$ is a Lie ideal of $K$, and $[U, K] \subset U$.

If $x \in R$, $x = s + k$ where $s \in S$, $k \in K$ thus $ux + x^*u = us + su + uk - ku \in U$ for all $u \in U$. Therefore, $(ux + x^*u)y + y^*(ux + x^*u) \in U$; this gives $x^*uy + y^*ux \in U$ for all $x, y \in R$. If $r \in R$ then $r = \sum x_i u y_i$ if $u \neq 0 \in U$; thus $r^* = -\sum y_i^* u x_i^*$ and so $r - r^* = \sum (x_i u y_i + y_i^* u x_i^*) \in U$. But since char $R \neq 2$, every $k \in K$ is of the form $r - r^*$. Hence $U = K$ follows.

We close this section with a result which generalizes that of the preceding theorem. In it we characterize all additive subgroups, $T$, of $R$ such that $S \circ T \subset T$.

THEOREM 4.1.7. Let $R$ be a simple ring, of characteristic not 2, with involution, such that $\dim_Z R > 4$. Let $T$ be an additive subgroup of $R$ such that $S \circ T \subset T$. If the involution (on $\mathfrak{z}(R)$) is of the first kind then $T = 0$, $S, K$, or $R$. If the involution is of the second kind then $T = 0$, $\lambda S$, or $R$ where $\lambda$ is in the centroid, $\mathfrak{z}(R)$, of $R$.

<u>Proof.</u> If $S \cap T \neq 0$ then $S \cap T$ is a Jordan ideal of $S$, hence must equal S by Theorem 2.1.12. In this case, if $T \neq S$, then we get that $K \cap T \neq 0$ and $S \circ (K \cap T) \subset K \cap T$. By Theorem 4.1.6 we have

$K \cap T = K$, whence $T \supset S + K = R$. Similarly, if $K \cap T \neq 0$ and $K \neq T$ we get $T = R$.

Suppose, then, that $S \cap T = K \cap T = 0$. If $0 \neq t \in T$ we can write $t = s + k$ where $s \in S$, $k \in K$. The set of such $s$ forms a Jordan ideal of S, so is all of S, since $S \circ T \subset T$. Thus for every $s \in S$ there exists a $k \in K$, which we write as $\theta(s)$, such that $s + \theta(s) \in T$. Since $S \cap T = 0$, $\theta(s)$ is unique, so $\theta$ is a well-defined additive map from S into K. Moreover, as is trivial to verify, $\theta(as + sa) = a\theta(s) + \theta(s)a$ for all $a, s \in S$. Consequently, by Theorem 4.1.4, $\theta(s) = \lambda s$ for all $s \in S$ where $\lambda \in \mathfrak{Z}(R)$. Hence $t = s + \lambda s = (1 + \lambda)s$. In short, $T = (1 + \lambda)S$. The theorem is thereby proved.

One remark should be made here. Although the mapping results were proved for simple rings they have analogous extensions to prime rings, and often to semi-prime rings.

## 2. Theorems of Martindale

We shall now turn to the exposition of several well-known results due to Martindale [4]. These results play an important role in the theory of Jordan Algebras. To see how the principal one of these theorems, Theorem 4.2.2, is exploited there the reader should look at Jacobson's book on Jordan Algebras [2].

Our concern will be with Jordan-type maps — either homomorphisms or derivations — on the set of symmetric elements in an appropriate class. The objective will be to show that these mappings come from restriction of the analogous associative map on all of the ring.

The results are motivated by a theorem of Jacobson and Rickart [3] which asserts: let $R = A_n$ be a ring with involution * such that $e_{ij}^* = e_{ji}$ for all the matrix units $e_{ij}$. Suppose that $n \geq 3$ and that every $s \in S$ is a trace, that is, $s = x + x^*$ for suitable $x \in R$. If $\varphi: S \to R'$ is an additive map such that $\varphi(s^2) = \varphi(s)^2$ and $\varphi(sts) = \varphi(s)\varphi(t)\varphi(s)$, for every $s, t \in S$ then $\varphi$ can be extended to an associative homomorphism of $R$ into $R'$.

DEFINITION. A mapping $\varphi: S \to R'$ which is additive and satisfies $\varphi(s^2) = \varphi(s)^2$ and $\varphi(sts) = \varphi(s)\varphi(t)\varphi(s)$ for all $s, t \in S$ is called a Jordan homomorphism of $S$ into $R'$.

Note that if $R'$ is 2-torsion free then it follows easily, from $\varphi(s^2) = \varphi(s)^2$, that $\varphi(sts) = \varphi(s)\varphi(t)\varphi(s)$, so in this case the second multiplicative condition imposed in the definition of Jordan homomorphism is redundant.

Note, too, that the assumption used by Lynne Small in Theorem 4.1.1, namely that $\varphi(xx^*) = \varphi(x)\varphi(x^*)$, is stronger than that used in the definition of Jordan homomorphism. One would thus expect that her result would be a consequence of the theorem we are about to prove. This is not the case, for while she imposes conditions on the nature of the subring of $R'$ generated by $\varphi(R)$, she has no condition requiring the existence of idempotents. In Martindale's theorem we require the existence of appropriate idempotents.

The proof will be a painstaking piece-by-piece extension of our given mapping to larger and larger pieces of the ring, showing at each such stage that the extension constructed is well-behaved. We now turn to the matter in its full detail.

In what follows $\varphi$ will be a Jordan homomorphism of $S$ into $R'$. $R$ will be a ring with 1.

LEMMA 4.2.1. If $e$ is a symmetric idempotent in $R$ then $\varphi(e)$ is an idempotent.

Proof. Trivial, since $\varphi(e) = \varphi(e^2) = \varphi(e)^2$, because $e = e^* \in S$.

LEMMA 4.2.2. If $e_i, e_j$ are symmetric orthogonal (i.e., $e_ie_j = e_je_i = 0$) in $R$ then $\varphi(e_i)$ and $\varphi(e_j)$ are orthogonal idempotents in $R'$.

Proof. That $\varphi(e_i)$ and $\varphi(e_j)$ are idempotents we already know by the previous lemma. Why orthogonal? For, since $\varphi(e_i)\varphi(e_j)\varphi(e_i) = \varphi(e_ie_je_i) = \varphi(0) = 0$, we have

$$\begin{aligned}\varphi(e_i)\varphi(e_j) &= \varphi(e_i)\varphi(e_i)\varphi(e_j)\\ &= \varphi(e_i)\varphi(e_i)\varphi(e_j) + \varphi(e_i)\varphi(e_j)\varphi(e_i)\\ &= \varphi(e_i)(\varphi(e_ie_j + e_je_i)) = 0\end{aligned}$$

since $e_ie_j + e_je_i = 0$.

DEFINITION. We say that $R$ satisfies Condition $(A_n)$, $n > 1$, if $R$ has n non-zero orthogonal symmetric idempotents whose sum is 1.

Suppose that $R$ satisfies Condition $(A_n)$ for some $n > 1$. By the Peirce decomposition, $R = \sum_{i,j} R_{ij}$ where $R_{ij} = e_iR_j$. Therefore $R_{ij}^* = R_{ji}$. We use the notation: if $x_{ij} \in R_{ij}$ then $x_{ji} = x_{ij}^*$.

If $s \in S$ then $s = \sum_{i,j} e_ise_j = \sum_i e_ise_i + \sum_{i<j} (e_ise_j + e_jse_i) = \sum s_i + \sum_{i<j} (x_{ij} + x_{ji})$ where $s_i \in S_i = S \cap R_{ii}$.

Let $R'_{ij} = \varphi(e_i)R'\varphi(e_j)$. In view of Lemma 4.2.2, $R'_{ij}R'_{k\ell} = 0$ if $j \neq k$.

LEMMA 4.2.3. $\varphi(x_{ij} + x_{ji}) \in R'_{ij} + R'_{ji}$.

Proof. $\varphi(x_{ij} + x_{ji}) = \varphi(e_i(x_{ij} + x_{ji})e_j + e_j(x_{ij} + x_{ji})e_i)$

$$= \varphi(e_i)\varphi(x_{ij} + x_{ji})\varphi(e_j) + \varphi(e_j)\varphi(x_{ij} + x_{ji})\varphi(e_i)$$

is in $R'_{ij} + R'_{ji}$. Here we have made use of the linearized form of $\varphi(sts) = \varphi(s)\varphi(t)\varphi(s)$.

DEFINITION. $\psi: \sum_{i \neq j} R_{ij} \to R'$ is defined by

$$\psi(x_{ij}) = \varphi(e_i)\varphi(x_{ij} + x_{ji})\varphi(e_j).$$

We now look at some properties of $\psi$. Clearly, from its form, $\psi(R_{ij}) \subset R'_{ij}$. Also,

$$\begin{aligned} \psi(x_{ij} + x_{ji}) &= \psi(x_{ij}) + \psi(x_{ji}) \\ &= \varphi(e_i)\varphi(x_{ij} + x_{ji})\varphi(e_j) + \varphi(e_j)\varphi(x_{ij} + x_{ji})\varphi(e_i) \\ &= \varphi(e_i x_{ij} e_j + e_i x_{ji} e_j + e_j x_{ij} e_i + e_j x_{ji} e_i) \\ &= \varphi(x_{ij} + x_{ji}), \end{aligned}$$

from the linearized form of $\varphi(sts) = \varphi(s)\varphi(t)\varphi(s)$. We shall use this last identity, between $\varphi$ and $\psi$, consistently.

LEMMA 4.2.4. If $i \neq j$, then $\varphi(x_{ij}y_{ji} + y_{ij}x_{ji}) = \psi(x_{ij})\psi(y_{ji}) + \psi(y_{ij})\psi(x_{ji})$.

Proof. $\varphi(x_{ij}y_{ji} + y_{ij}x_{ji}) + \varphi(x_{ji}y_{ij} + y_{ji}x_{ij})$

$$\begin{aligned} &= \varphi\,\{(x_{ij} + x_{ji})(y_{ij} + y_{ji}) + (y_{ij} + y_{ji})(x_{ij} + x_{ji})\} \\ &= \varphi(x_{ij} + x_{ji})\varphi(y_{ij} + y_{ji}) + \varphi(y_{ij} + y_{ji})\,\varphi(x_{ij} + x_{ji}) \\ &= \psi(x_{ij} + x_{ji})\psi(y_{ij} + y_{ji}) + \psi(y_{ij} + y_{ji})\psi(x_{ij} + x_{ji}) \\ &= \psi(x_{ij})\psi(y_{ji}) + \psi(y_{ji})\psi(x_{ij}) + \psi(y_{ij})\psi(x_{ji}) + \psi(x_{ji})\psi(y_{ij}). \end{aligned}$$

From the definition of $\psi$, $\psi(x_{ij})\psi(y_{ji}) \in R'_{ii}$, and we know that the sum of the $R'_{ij}$'s is direct (Lemma 4.2.2). Also, $\varphi(S \cap R_{ii}) \subset R'_{ii}$. Thus, equating the $R'_{ii}$ component on the left and right sides of the long chain of equalities above yields the required result, namely, that $\varphi(x_{ij}y_{ji} + y_{ij}x_{ji}) = \psi(x_{ij})\psi(y_{ji}) + \psi(y_{ij})\psi(x_{ji})$.

If $s = \sum_\sigma x^\sigma_{ij} y^\sigma_{ji} \in S$, and so, in $S \cap R_{ii}$, and if $i \neq j$, applying * we get, on addition, that $2s = \sum_\sigma (x^\sigma_{ij} y^\sigma_{ji} + y^\sigma_{ij} x^\sigma_{ji})$. Using the preceding lemma we accordingly obtain

LEMMA 4.2.5. If $i \neq j$ and $s = \sum x^\sigma_{ij} y^\sigma_{ji}$ is in $S$, then $2\varphi(s) = \sum \{\psi(x^\sigma_{ij})\psi(y^\sigma_{ji}) + \psi(y^\sigma_{ij})\psi(x^\sigma_{ji})\}$ is in $\psi(R_{ij})\psi(R_{ji})$.

We continue with the investigation of the nature of $\psi$.

LEMMA 4.2.6. If $s_i \in S_i = S \cap R_{ii}$ and, for $i \neq j$, $x_{ij} \in R_{ij}$ then $\psi(s_i x_{ij}) = \varphi(s_i)\psi(x_{ij})$ and $\psi(x_{ji}s_i) = \psi(x_{ji})\varphi(s_i)$.

<u>Proof.</u>
$$\begin{aligned}\varphi(s_i x_{ij} + x_{ji}s) &= \varphi(s_i(x_{ij} + x_{ji}) + (x_{ij} + x_{ji})s_i) \\ &= \varphi(s_i)\{\psi(x_{ij}) + \psi(x_{ji})\} + \{\psi(x_{ij}) + \psi(x_{ji})\}\varphi(s_i) \\ &= \varphi(x_i)\psi(x_{ij}) + \psi(x_{ji})\varphi(s_i)\end{aligned}$$

since $\varphi(s_i) \in R'_{ii}$ and $\psi(x_{ji}) \in R'_{ji}$. Since $\varphi(s_i x_{ij} + x_{ji}s_i) = \psi(s_i x_{ij}) + \psi(x_{ji}s_i)$, reading off the components in $R'_{ij}$ and $R'_{ji}$ gives us the results of the lemma.

LEMMA 4.2.7. If $i \neq j$ and $s_j \in S_j$, $x_{ij} \in R_{ij}$ then $\varphi(x_{ij}s_j x_{ji}) = \psi(x_{ij})\varphi(s_j)\psi(x_{ji})$.

<u>Proof.</u> $s_j \in S_j$, $x_{ij} \in R_{ij}$ and $i \neq j$, $x_{ij}s_j x_{ji} = (x_{ij} + x_{ji})s_j(x_{ij} + x_{ji})$. Hence
$$\begin{aligned}\varphi(x_{ij}s_j x_{ji}) &= \varphi((x_{ij} + x_{ji})s_j(x_{ij} + x_{ji})) = \varphi(x_{ij} + x_{ji})\varphi(s_j)\varphi(x_{ij} + x_{ji}) \\ &= \psi(x_{ij} + x_{ji})\varphi(s_j)\psi(x_{ij} + x_{ji}) = \psi(x_{ij})\varphi(s_j)\psi(x_{ji}),\end{aligned}$$

from the orthogonalities that we know, by Lemma 4.2.2.

DEFINITION. If $R$ satisfies Condition $(A_n)$ then we say that it satisfies Condition $(B_n)$ if for all $i, j$ we have $R_{ij}R_{ji} = R_{ii}$.

It is easy to see that $R$ satisfies Condition $(B_n)$ if and only if $Re_iR = R$ for $i = 1, 2, \ldots, n$.

LEMMA 4.2.8. If $R$ satisfies Conditions $(A_n)$ and $(B_n)$ for some $n > 1$ and if $\varphi$ can be extended to an associative homomorphism $\Phi$ of $R$ into $R'$, then $\Phi$ is unique.

Proof. By Condition $(B_n)$, $R$ is generated by its elements $x_{ij}$ with $i \neq j$. But $\Phi(x_{ij}) = \Phi(e_i(x_{ij} + x_{ji})) = \Phi(e_i)\Phi(x_{ij} + x_{ji}) = \varphi(e_i)\varphi(x_{ij} + x_{ji})$. Thus $\Phi(x_{ij})$ is uniquely determined. Since the $x_{ij}$ generate $R$ we have that $\Phi$ is uniquely determined on $R$.

We assume through the sequence of lemmas to come now that $R$ satisfies Conditions $(A_3)$ and $(B_3)$.

We show an important multiplicative property of the mapping $\psi$.

LEMMA 4.2.9. If $i, j, k$ are distinct then $\psi(x_{ij}x_{jk}) = \psi(x_{ij})\psi(x_{jk})$.

Proof. $\psi(x_{ij}x_{jk}) + \psi(x_{kj}x_{ji}) = \varphi(x_{ij}x_{jk} + x_{kj}x_{ji})$

$$= \varphi((x_{ij} + x_{ji})(x_{jk} + x_{kj}) + (x_{jk} + x_{kj})(x_{ij} + x_{ji}))$$

$$= \psi(x_{ij} + x_{ji})\psi(x_{jk} + x_{kj})\psi(x_{ij} + x_{ji}).$$

Expanding and reading off the component in $R'_{ik}$ gives us the result of the lemma.

LEMMA 4.2.10. If $i \neq j$ and $s_i = \sum_\sigma x^\sigma_{ij} y^\sigma_{ji} \in S_i$, then for $k \neq q$, $\{\varphi(s_i) - \sum_\sigma \psi(x^\sigma_{ij})\psi(y^\sigma_{ji})\}\psi(R_{kq}) = 0$.

<u>Proof.</u> If $k \neq i$ the result is trivial by the previous lemmas. So, suppose $i = k = 1$, $j = 2$.

If $q = 2$ then $e_1 = \sum_\mu a_{13}^\mu b_{31}^\mu$, since by Condition $(B_n)$, $R_{11} = R_{13}R_{31}$. Let $c_{12} \in R_{12}$. We make a long computation. By Lemmas 4.2.6 and 4.2.9, used repeatedly,

$$\begin{aligned}
\varphi(s_1)\psi(c_{12}) = \psi(s_1 c_{12}) &= \psi(\sum_\sigma x_{12}^\sigma y_{21}^\sigma c_{12}) = \sum_\sigma \psi(x_{12}^\sigma y_{21}^\sigma e_1 c_{12}) \\
&= \sum_{\sigma,\mu} \psi(x_{12}^\sigma y_{21}^\sigma a_{13}^\mu b_{31}^\mu c_{12}) = \sum_{\sigma,\mu} \psi(x_{12}^\sigma y_{21}^\sigma a_{13}^\mu)\psi(b_{31}^\mu c_{12}) \\
&= \sum_{\sigma,\mu} \psi(x_{12}^\sigma)\psi(y_{21}^\sigma a_{13}^\mu)\psi(b_{31}^\mu c_{12}) \\
&= \sum_{\sigma,\mu} \psi(x_{12}^\sigma)\psi(y_{21}^\sigma)\psi(a_{13}^\mu)\psi(b_{31}^\mu c_{12}) \\
&= \sum_{\sigma,\mu} \psi(x_{12}^\sigma)\psi(y_{21}^\sigma)\psi(a_{13}^\mu b_{31}^\mu c_{12}) \\
&= \sum_{\sigma} \psi(x_{12}^\sigma)\psi(y_{21}^\sigma)\psi(e_1 c_{12}) \\
&= \sum_{\sigma} \psi(x_{12}^\sigma)\psi(y_{21}^\sigma)\psi(c_{12}) .
\end{aligned}$$

Hence $(\varphi(s_1) - \sum_\sigma \psi(x_{12}^\sigma)\psi(y_{21}^\sigma))\psi(R_{12}) = 0$ .

On the other hand, if $q \neq 2$ we can suppose $q = 3$; then

$$\begin{aligned}
\varphi(s_1)\psi(c_{13}) = \psi(s_1 c_{13}) &= \sum_\sigma \psi(x_{12}^\sigma y_{21}^\sigma c_{13}) \\
&= \sum_\sigma \psi(x_{12}^\sigma)\psi(y_{21}^\sigma c_{13}) \quad \text{by Lemma 4.2.9} \\
&= \sum_\sigma \psi(x_{12}^\sigma)\psi(y_{21}^\sigma)\psi(c_{13}) \quad \text{by Lemma 4.2.9.}
\end{aligned}$$

From this we obtain the required result

$$(\varphi(s_1) - \sum_\sigma \psi(x_{12}^\sigma)\psi(y_{21}^\sigma))\psi(R_{13}) = 0.$$

So far we have the mapping $\psi$ defined, however it is only defined on the pieces $R_{ij}$ where $i \neq j$. We want a mapping on all of R, hence on the pieces $R_{ii}$. Now $R_{ii} = R_{ij}R_{ji}$ if $j \neq i$ by Condition $(B_n)$. To extend $\psi$ to the $R_{ii}$'s now requires some kind of uniqueness. This vague statement is made precise in the next lemma.

LEMMA 4.2.11. If $\sum_{\sigma} x_{ij}^{\sigma} y_{ji}^{\sigma} = \sum_{\mu} v_{ik}^{\mu} w_{ki}^{\mu}$ where i, j and k are distinct, then

$$\sum_{\sigma} \psi(x_{ij}^{\sigma})\psi(y_{ji}^{\sigma}) = \sum_{\mu} \psi(v_{ik}^{\mu})\psi(w_{ki}^{\mu}) .$$

Proof. Let i, j, k be 1, 2, 3 respectively and let $c_{13} \in R_{13}$, $c_{21} \in R_{21}$. Using Lemma 4.2.9 repeatedly, we have

$$\begin{aligned}
\psi(c_{21}) \sum_{\sigma} \psi(x_{12}^{\sigma})\psi(y_{21}^{\sigma})\psi(c_{13}) &= \psi(c_{21}) \sum \psi(x_{12}^{\sigma})\psi(y_{21}^{\sigma}c_{13}) \\
&= \psi(c_{21}) \sum \psi(x_{12}^{\sigma}y_{21}^{\sigma}c_{13}) \\
&= \psi(c_{21})\psi(\sum_{\sigma} x_{12}^{\sigma}y_{21}^{\sigma}c_{13}) \\
&= \psi(c_{21})\psi(\sum_{\mu} v_{13}^{\mu}w_{31}^{\mu}c_{13}) \\
&= \psi(\sum c_{21}v_{13}^{\mu}w_{31}^{\mu}c_{13}) .
\end{aligned}$$

Now unravel this last term by pulling $\psi(c_{13})$ out on the right and proceeding in a manner similar to the one just used. We get

$$\psi(c_{21})\{\sum_{\sigma} \psi(x_{12}^{\sigma})\psi(y_{21}^{\sigma}) - \sum_{\mu} \psi(v_{13}^{\mu})\psi(w_{31}^{\mu})\}\psi(c_{13}) = 0.$$ If

$$u' = \sum_{\sigma} \psi(x_{12}^{\sigma})\psi(y_{21}^{\sigma}) - \sum_{\mu} \psi(v_{13}^{\mu})\psi(w_{31}^{\mu})$$ then $u' \in R'_{11}$ and $\psi(R_{21})u'(R_{13}) = 0$. We want to show that $u' = 0$.

Also, $e_1 = \sum_\sigma a^\sigma_{12} b^\sigma_{21}$ and, by Lemma 4.2.10

$$(1) \quad \{\varphi(e_1) - \sum_\sigma \psi(a^\sigma_{12})\psi(b^\sigma_{21})\}\psi(R_{1m}) = 0 \quad \text{if } m = 2 \text{ or } 3.$$

Now

$$u' = \varphi(e_1)u'\varphi(e_1) = \{\varphi(e_1) - \sum_\sigma \psi(a^\sigma_{12})\psi(b^\sigma_{21})\}u'\varphi(e_1)$$
$$+ \sum_\sigma \psi(a^\sigma_{12})\psi(b^\sigma_{21})u'\{\varphi(e_1) - \sum_\mu \psi(a^\mu_{13})\psi(b^\mu_{31})\}$$
$$+ \sum_\sigma \psi(a^\sigma_{12})\psi(b^\sigma_{21})\, u' \sum_\mu \psi(a^\mu_{13})\psi(b^\mu_{31})$$

where $e_1 = \sum_\mu a^\mu_{13} b^\mu_{31}$ (since $R_{11} = R_{13}R_{31}$).

The last term is 0 since we have $\psi(R_{21})u'\psi(R_{13}) = 0$. We claim that the first two terms are also 0. For, what does $u'$ look like? It is in $\psi(R_{12})\psi(R_{21}) + \psi(R_{13})\psi(R_{31})$ and, by (1), $\varphi(e_1) - \sum_\sigma \psi(a^\sigma_{12})\psi(b^\sigma_{21})$ annihilates $\psi(R_{12})$ and $\psi(R_{13})$ from the left. This accounts for the first term. The analog of (1) gives that $\psi(R_{21})$ and $\psi(R_{31})$ are annihilated on the right by $\varphi(e_1) - \sum_\mu \psi(a^\mu_{13})\psi(b^\mu_{31})$. This accounts for the second term. Hence $u' = 0$, and the lemma is proved.

With the uniqueness given us by Lemma 4.2.11 we are able to define a mapping on all of R.

DEFINITION. $\Phi: R \to R$ is defined by

(1) $\Phi(x_{ij}) = \psi(x_{ij}) \in R'_{ij}$ if $i \neq j$,

(2) $\Phi(x_{ii}) = \sum_\sigma \psi(x^\sigma_{ij})\psi(y^\sigma_{ji}) \in R'_{ii}$ if $x_{ii} = \sum_\sigma x^\sigma_{ij} y^\sigma_{ji}$ (for some $j \neq i$).

Lemma 4.2.11 assures us that the mapping $\Phi$ is well-defined.

We are finally able to prove the first of our theorems.

THEOREM 4.2.1. $\Phi$ is a homomorphism of R into R'.

Proof. To prove the result we must verify the following:

a) $\Phi(x_{ij}x_{k\ell}) = 0 = \Phi(x_{ij})\Phi(y_{k\ell})$ if $j \neq k$.

b) $\Phi(x_{12}x_{23}) = \Phi(x_{12})\Phi(x_{23})$ (we are using 1,2,3 as any three distinct indices). By Lemma 4.2.9, $\psi(x_{12}x_{23}) = \psi(x_{12})\psi(x_{23})$, and since $\Phi(x_{12}x_{23}) = \psi(x_{12}x_{23})$, $\Phi(x_{12}) = \psi(x_{12})$ and $\Phi(x_{23}) = \psi(x_{23})$, from the definition of $\Phi$, b) is valid.

c) $\Phi(x_{12}y_{21}) = \Phi(x_{12})\Phi(y_{21})$.

By the definition of $\Phi$, $\Phi(x_{12}y_{21}) = \psi(x_{12})\psi(y_{21}) = \Phi(x_{12})\Phi(y_{21})$; hence c) holds.

d) $\Phi(x_{11}y_{12}) = \Phi(x_{11})\Phi(y_{12})$.

Since $R_{11} = R_{13}R_{31}$, $x_{11} = \sum_{\sigma} x_{12}^{\sigma} y_{31}^{\sigma}$, whence $x_{11}y_{12} = \sum_{\sigma} x_{13}^{\sigma} y_{31}^{\sigma} y_{12}$. Using Lemma 4.2.9 twice on $\Phi(\sum_{\sigma} x_{13}^{\sigma} y_{31}^{\sigma} y_{12})$ we get

$$\Phi(x_{11}y_{12}) = \sum_{\sigma} \psi(x_{13}^{\sigma})\psi(y_{31}^{\sigma})\psi(y_{12}) = \Phi(x_{11})\Phi(y_{12}),$$

from the definition of $\Phi$.

e) $\Phi(x_{11}y_{11}) = \Phi(x_{11})\Phi(y_{11})$.

Now $x_{11} = \sum_{\sigma} x_{12}^{\sigma} y_{21}^{\sigma}$, hence

$$\begin{aligned}\Phi(x_{11}y_{11}) &= \sum_{\sigma} \Phi(x_{12}^{\sigma} y_{21}^{\sigma} y_{11}) = \sum_{\sigma} \Phi(x_{12}^{\sigma})\Phi(y_{21}^{\sigma} y_{11}) \quad \text{by part c)} \\ &= \sum_{\sigma} \Phi(x_{12}^{\sigma})\Phi(y_{21}^{\sigma})\Phi(y_{11}) \quad \text{by part d)} \\ &= \sum \Phi(x_{12}^{\sigma} y_{21}^{\sigma})\Phi(y_{11}) \quad \text{by part c)} \\ &= \Phi(x_{11})\Phi(y_{11}), \quad \text{as desired.}\end{aligned}$$

Therefore $\Phi$ has been shown to be a homomorphism. This proves the theorem.

Now that we know that $\Phi$ is a homomorphism, it is natural to wonder: when is $\Phi$ an extension of $\varphi$? The requisite conditions turn out to be as in

LEMMA 4.2.12. If either $\varphi(S)$ is 2-torsion free, or if $S = T = \{x + x^* \mid x \in R\}$, then $\Phi$ is an extension of $\varphi$.

Proof. By the definition of $\Phi$, if $i \neq j$ then

$$\Phi(x_{ij} + x_{ji}) = \psi(x_{ij}) + \psi(x_{ji}) = \varphi(x_{ij} + x_{ji}).$$

As we saw earlier, just prior to Lemma 4.2.3, if $s \in S$ then $s = \sum_i s_i + \sum_{i<j} (x_{ij} + x_{ji})$. Thus, to show that $\Phi(s) = \varphi(s)$, we merely must show that $\Phi(s_i) = \varphi(s_i)$ where $s_i \in S_i = S \cap R_{ii}$.

If $\varphi(S)$ is 2-torsion free, since $s_i = \sum_\sigma x_{ij}^\sigma y_{ji}^\sigma$ for $j \neq i$, by Lemma 4.2.5

$$2\varphi(s_i) = \sum_\sigma \{\psi(x_{ij}^\sigma)\psi(y_{ji}^\sigma) + \psi(y_{ij}^\sigma)\psi(x_{ji}^\sigma)\}$$

$$= 2\Phi(s_i) \quad \text{(from the definition of } \Phi\text{)}.$$

Since $\varphi(S)$ is 2-torsion free, we get $\varphi(s_i) = \Phi(s_i)$.

On the other hand, if $S = T = \{ x + x^* \mid x \in R\}$, then

$$s_i = x_i + x_i^* = \sum_\sigma x_{ij}^\sigma y_{ji}^\sigma + \sum_\sigma y_{ij}^\sigma x_{ji}^\sigma$$

for $j \neq i$. Thus, by Lemma 4. 2. 4,

$$\varphi(s_i) = \sum_\sigma \{\psi(x_{ij}^\sigma)\psi(y_{ji}^\sigma) + \psi(y_{ij}^\sigma)\psi(x_{ji}^\sigma) = \Phi(s_i) .$$

The lemma is proved.

We now have all the pieces to prove the important result due to Martindale [4].

THEOREM 4.2.2. Let R be a ring with involution $*$, having non-zero orthogonal symmetric idempotents $e_1, e_2, e_3$ such that $e_1 + e_2 + e_3 = 1$, the unit element of R. Suppose that $Re_iR = R$ for $i = 1, 2, 3$. If $\varphi$ is a Jordan homomorphism of S into R' such that either $\varphi(S)$ is 2-torsion free or $S = T = \{x+x^* \mid x \in R\}$ then $\varphi$ can be extended in a unique way to an associative homomorphism of R into R'.

COROLLARY 1. Let R be a simple ring with 1, and $*$ an involution on R such that $S = T = \{x+x^* \mid x \in R\}$. If R has three non-zero orthogonal symmetric idempotents such that $e_1 + e_2 + e_3 = 1$, then any non-zero Jordan homomorphism of S into R' can be extended to an isomorphism of R into R'.

Proof. If R is simple, and e is an idempotent in R, then $ReR = R$ follows automatically. Thus the conditions of the theorem are fulfilled, hence the corollary follows.

Note that if char $R \neq 2$, in this corollary, then $S = T$ is a consequence. Hence the corollary holds for simple rings of characteristic not 2.

DEFINITION. If R is a ring with involution, the additive mapping $\delta: S \to R$ is a Jordan derivation of S into R if $\delta(s^2) = s\delta(s) + \delta(s)s$ for every $s \in S$, and $\delta(sts) = \delta(s)ts + s\delta(t)s + st\delta(s)$ for $s, t \in S$.

COROLLARY 2. If R is as in Theorem 4.2.2 and $\delta$ is a Jordan derivation of S into R then $\delta$ can be extended to a derivation of R.

Proof. Let $R' = \left\{\begin{pmatrix} a & b \\ 0 & c \end{pmatrix} \,\middle|\, a,b,c \in R\right\}$; define $\varphi: S \to R'$ by $\varphi(s) = \begin{pmatrix} s & \delta(s) \\ 0 & s \end{pmatrix}$. Then, trivially, $\varphi(s^2) = \varphi(s)^2$ and $\varphi(sts) = \varphi(s)\varphi(t)\varphi(s)$ for $s,t \in S$. By the theorem, $\varphi$ can be extended to a homomorphism $\Phi$ of $R$ into $R'$. If we write

$$\Phi(x) = \begin{pmatrix} f(x) & D(x) \\ 0 & g(x) \end{pmatrix}$$

, for $x \in R$, then $f$ is a homomorphism of $R$ into $R$, which is the identity map on $S$. By Lemma 4.2.8, $f(x) = x$ for all $x \in R$. Similarly, $g(x) = x$ for all $x \in R$. Then $D$ must be a derivation of $R$ into $R$, and since $D(s) = \delta(s)$, for $s \in S$, $D$ is an extension of $\delta$.

We now want to look at the situation where $R$ has a symmetric idempotent $e \neq 0, 1$ but does not have three orthogonal symmetric idempotents whose sum is 1. We do assume, however, that $ReR = R$ and $R(1-e)R = R$. Let $e_1 = e$ and $e_2 = 1-e$.

We add a further hypothesis on $R$.

<u>Condition (C)</u>: For $i = 1, 2$, $S_i = S \cap e_iRe_i$ <u>generates</u> $e_iRe_i = R_{ii}$.

We suppose, also, that $R'$ is 2-torsion free.

In Lemma 4.2.6 we showed that $\psi(s_ix_{ij}) = \varphi(s_i)\psi(x_{ij})$ if $i \neq j$ and $s_i \in S_i$, $x_{ij} \in R_{ij}$. Iterating the use of Lemma 4.2.6 we clearly get, for $s_i^{(1)}, \ldots, s_i^{(k)} \in S_i$, $x_{ij} \in R_{ij}$ that

$$\psi(s_i^{(1)} \cdots s_i^{(k)} x_{ij}) = \varphi(s_i^{(1)}) \cdots \varphi(s_i^{(k)})\psi(x_{ij}) .$$

This observation enables us to prove

LEMMA 4.2.13. If $s_i^\sigma, t_i^\sigma, \ldots, v_i^\sigma \in S_i$ and $\sum_\sigma s_i^\sigma t_i^\sigma \cdots v_i^\sigma = 0$ then $\sum_\sigma \varphi(s_i^\sigma)\varphi(t_i^\sigma) \cdots \varphi(v_i^\sigma) = 0$.

Proof. By the remark above, if $x_{ij} \in R_{ij}$ where $j \neq i$ then

$$0 = \psi(\sum_\sigma s_i^\sigma t_i^\sigma \cdots v_i^\sigma x_{ij}) = \sum_\sigma \varphi(s_i^\sigma)\varphi(t_i^\sigma)\cdots \varphi(v_i^\sigma)\psi(x_{ij}),$$

hence

$$\sum_\sigma \varphi(s_i^\sigma)\varphi(t_i^\sigma)\cdots \varphi(v_i^\sigma))\psi(R_{ij}) = 0.$$

Now $e_i \in S_i \subset R_{ii} = R_{ij}R_{ji}$, hence $e_i = \sum_\sigma x_{ij}^\sigma y_{ji}^\sigma$. By Lemma 4.2.5, $2\varphi(e_i) \in \psi(R_{ij})\psi(R_{ji})$, hence

$$2\{\sum \varphi(s_i^\sigma)\cdots \varphi(v_i^\sigma)\}\varphi(e_i) = 0 .$$

But $\varphi(e_i)$ is the unit element of $\varphi(R_{ii})$ and $\sum_\sigma \varphi(s_i^\sigma)\cdots\varphi(v_i^\sigma)$ is in $\varphi(R_{ii})$. Together with the fact that $R'$ is 2-torsion free, this leads to the conclusion of the lemma.

DEFINITION. $\Phi: R \to R'$ is defined by

(1) $\Phi(x_{ij}) = \psi(x_{ij})$ if $i \neq j$

(2) if $x_i \in R_{ii}$ and, since $S_i$ generates $R_{ii}$, $x_i = \sum s_i^\sigma \cdots v_i^\sigma$, $s_i^\sigma, \ldots, v_i^\sigma \in S_i$ then $\Phi(x_i) = \sum_\sigma \varphi(s_i^\sigma)\cdots\varphi(v_i^\sigma)$.

By Lemma 4.2.13, $\Phi$ is well-defined on $R$. It trivially is an additive mapping extending $\varphi$ to $R$.

THEOREM 4.2.3. Let $R$ be a ring with involution, having an idempotent $e = e^* \neq 0, 1$ such that $ReR = R(1-e)R = R$ and such that Condition (C) is satisfied. Let $\varphi$ be a Jordan homomorphism of S into $R'$, where $R'$ is 2-torsion free. Then $\varphi$ can be uniquely extended to an associative homomorphism of $R$ into $R'$.

Proof. Exactly as we showed in Lemma 4.2.8, if $\varphi$ can be extended to a homorphism of R it can only be done so in one way. We now show that $\Phi$ is the required homomorphism extension of $\varphi$.

a) If $x_i, y_i \in R_{ii}$ then $\Phi(x_i y_i) = \Phi(x_i)\Phi(y_i)$. This is clear from the definition of $\Phi$ on $R_{ii}$ and the fact that $S_i$ generates $R_{ii}$.

b) If $x_i \in R_{ii}$, $y_{ij} \in R_{ij}$ where $j \neq i$ then $\Phi(x_i y_{ij}) = \Phi(x_i)\Phi(y_{ij})$. This follows from the remark just preceding Lemma 4.2.13 and the fact that $S_i$ generates $R_{ii}$.

c) If $i \neq j$ then $\Phi(x_{ij} y_{ji}) = \Phi(x_{ij})\Phi(y_{ji})$. This is the only non-trivial part that needs a real verification.

Now $e_i = \sum_\sigma a_{ij}^\sigma b_{ji}^\sigma = \sum_\sigma b_{ij}^\sigma a_{ji}^\sigma$ ; hence, by Lemma 4.2.5, $2\varphi(e_i) = \sum_\sigma \{\psi(a_{ij}^\sigma)\psi(b_{ji}^\sigma) + \psi(b_{ij}^\sigma)\psi(a_{ji}^\sigma)\}$. Therefore

$$
\begin{aligned}
2\Phi(x_{ij}y_{ji}) &= 2\Phi(x_{ij}y_{ji})\varphi(e_i) \\
&= \Phi(x_{ij}y_{ji})\sum_\sigma \{\psi(a_{ij}^\sigma)\psi(b_{ji}^\sigma) + \psi(b_{ij}^\sigma)\psi(a_{ji}^\sigma)\} \\
&= \sum_\sigma \Phi(x_{ij}y_{ji})\psi(a_{ij}^\sigma)\psi(b_{ji}^\sigma) + \sum_\sigma \Phi(x_{ij}y_{ji})\psi(b_{ij}^\sigma)\psi(a_{ji}^\sigma) \\
&= \sum_\sigma \Phi(x_{ij}y_{ji}a_{ij}^\sigma)\psi(b_{ji}^\sigma) + \sum_\sigma \Phi(x_{ij}y_{ji}b_{ij}^\sigma)\psi(a_{ji}^\sigma)
\end{aligned}
$$

by part b).

Now $y_{ji}a_{ij}^\sigma \in R_{jj}$, so it is of the form $y_{ji}a_{ij}^\sigma = \sum_\lambda s_j^\lambda \cdots v_j^\lambda$ where $s_j^\lambda, \ldots, v_j^\lambda \in S_j$. Substitute into the identity above, and use the remark preceding Lemma 4.2.13 to get

$$\Phi(x_{ij}y_{ji}a_{ij}^\sigma) = \sum \psi(x_{ij})\varphi(s_j^\lambda)\cdots\varphi(v_j^\lambda) .$$

We get a similar expression for $\Phi(x_{ij}y_{ji}b_{ij}^\sigma)$. Using these above we obtain

$$2\Phi(x_{ij}y_{ji}) = \sum_{\lambda,\sigma} \psi(x_{12})\varphi(s_j^{\lambda})\cdots\varphi(v_j^{\lambda})\psi(b_{ji}^{\sigma}) + \text{a similar term}$$

$$= \sum_{\lambda,\sigma} \psi(x_{ij})\psi(s_j^{\lambda}\cdots v_j^{\lambda}b_{ji}^{\sigma}) + \text{a similar term}$$

$$= \sum_{\sigma} \psi(x_{ij})\psi(y_{ji}a_{ij}^{\sigma}b_{ji}^{\sigma}) + \sum_{\sigma} \psi(x_{ij})\psi(y_{ji}b_{ij}^{\sigma}a_{ji}^{\sigma})$$

$$= \psi(x_{ij})\psi(y_{ji}e_i) + \psi(x_{ij})\psi(y_{ji}e_i)$$

$$= 2\psi(x_{ij})\psi(y_{ji}) .$$

Since $R'$ is 2-torsion free we obtain $\Phi(x_{ij})\Phi(y_{ji}) = \Phi(x_{ij}y_{ji})$.

We have shown that $\Phi$ is a homomorphism, hence we have proved the theorem.

An immediate consequence of this theorem is a very pretty result for simple rings.

THEOREM 4.2.4. Let $R$ be a simple ring with involution of characteristic not 2. Suppose that $\varphi \neq 0$ is a Jordan homomorphism of $S$ into $R'$, where $R'$ is 2-torsion free. If $\dim_Z R > 16$ then $\varphi$ can be extended in a unique way to an isomorphism of $R$ into $R'$.

<u>Proof.</u> Let $e_1 = e$, $e_2 = 1-e$ and $R_{ii} = e_iRe_i$, $S_i = R_{ii} \cap S$. If $S_i$ generates $R_{ii}$ for $i = 1, 2$ then, by Theorem 4.2.3, the result is true. So we may suppose that $S_1$, say, does not generate $R_{11}$. Since $R_{11}$ is simple, and $S_1$ does not generate $R_{11}$, by Theorem 2.1.6, $R_{11}$ must be at most 4-dimensional over its center. But this immediately tells us that $R$ has a minimal right ideal; since $R$ is simple and has a unit element, $R$ must be simple artinian.

Also, neither $R_{11}$ nor $R_{22}$ can have a non-trivial symmetric idempotent, otherwise $R$ will have three non-trivial orthogonal symmetric idempotents, and so the result will follow from Theorem 4.2.2. By Theorem 1.2.2 we have that each of $R_{ii}$ is a division ring or the $2 \times 2$ matrices over a field.

If $R_{11}$ is a field or the $2 \times 2$ matrices over a field, then $R \approx F_n$, where $F$ is a field, for some $n$. But $R_2 = (1-e)R(1-e)$ is either $F$ or $F_2$ by the comments above. Since $e$ and $1-e$ are thus of rank at most 2, we have that 1 is of rank at most 4. Hence $n \leq 4$. But this gives the contradiction $\dim_Z R \leq 16$.

On the other hand, if $R_{11} = eRe$ is a non-commutative division ring $\Delta$, by Wedderburn's theorem $R \approx \Delta_n$. Hence $R_{22} \approx \Delta_{n-1}$. But $R_{22}$, we know, is either a division ring or the $2 \times 2$ matrices over a field. Since $\Delta$ is not a field, $R_{22}$ cannot be a $2 \times 2$ matrix ring over a field, whence $R_{22}$ is a division ring. Thus $n-1 = 1$ and so $n = 2$. Since $\Delta$ is 4-dimensional over its center, we again get the contradiction that $\dim_Z R \leq 16$. With this the theorem is proved.

Using the $2 \times 2$ matrix trick to pass from derivations to homomorphisms that we used before we have

THEOREM 4.2.5. Let $R$ be a ring with involution having a unit element 1, and a symmetric idempotent $e \neq 0, 1$. If $e_1 = e$ and $e_2 = 1-e$ satisfy $Re_iR = R$ and $S_i = e_iRe_i \cap S$ generates $e_iRe_i$ for $i = 1, 2$, then every Jordan derivation of $S$ into $R$ can be extended in a unique way to a derivation of $R$, if $R$ is 2-torsion free.

COROLLARY. If $R$ is a simple ring with involution of characteristic not 2 having a symmetric idempotent $e \neq 0, 1$ and such that

$\dim_Z R > 16$, then any Jordan derivation of $S$ into $R$ can be extended to a derivation of $R$ in a unique way.

We conclude these type of results with some examples to show, for simple ring, that these results may be false if $\dim_Z R \leq 16$.

Example 1. Let $F$ be the field of rational numbers and $R = F(x)$ the field of rational functions in $x$ over $F$. Define $*: R \to R$ by $f(x)^* = f(-x)$. This is an involution on $R$, and $S = F(x^2)$. The map $\varphi: S \to R$ is a Jordan homomorphism of $S$ into $R$. It clearly cannot be lifted to a homomorphism $\Phi$ of $R$ into $R$. For then $-x^2 = \Phi(x^2) = \Phi(x)^2$ would follow, and this is clearly false.

Of course, in Example 1, there are no non-trivial symmetric idempotents. We now give an example where there are such idempotents.

Example 2. Let $R = D_2$ be the $2 \times 2$ matrix ring over $D$, where $D$ is the division ring of quaternions over the reals. If $\bar{\alpha}$ denotes the ordinary involution of $D$ as applied to $\alpha \in D$, define $*: R \to R$ by $(\sum \alpha_{ij} e_{ij})^* = \sum \bar{\alpha}_{ij} e_{ji}$. This is an involution on $R$.

Define $\varphi: S \to R$ by:

$$\varphi(e_{11}) = e_{11}, \quad \varphi(e_{22} = e_{22}, \quad \varphi(i(e_{12} - e_{21})) = j(e_{12} - e_{21}),$$

$$\varphi(j(e_{12} - e_{21})) = i(e_{12} - e_{21}), \quad \varphi(k(e_{12} - e_{21})) = k(e_{12} - e_{21})$$

and

$$\varphi(e_{12} + e_{21}) = e_{12} + e_{21}.$$

This gives a Jordan homomorphism of $S$ into $R$. We leave it to the reader to show that $\varphi$ cannot be extended to a homomorphism (hence automorphism) of $R$.

The analogous types of questions for Lie homomorphisms of the skew-symmetric elements are of great interest and difficulty. The results, to date, are too long and complicated to include here. For the sharpest results in this direction the reader should look at the papers of Martindale [5, 6].

Bibliography

1. I. N. Herstein. Certain submodules in simple rings with involution. Duke Math. Jour. 24 (1957): 357-364.

2. N. Jacobson. Structure and Representation of Jordan Algebras. AMS. Colloq. Publ. XXXIX 1968.

3. N. Jacobson and C. Rickart. Homomorphisms of Jordan rings of self-adjoint elements. Trans. AMS 72 (1952): 310-322.

4. W. Martindale. Jordan homomorphisms of the symmetric elements of a ring with involution. Jour. Algebra 5 (1967): 232-249.

5. W. Martindale. Lie isomorphismss of the skew elements of a simple ring with involution. (To appear)

6. W. Martindale. Lie isomorphisms of the skew elements of a prime ring with involution. (To appear)

7. Lynne Small. Mappings on simple rings with involution. Jour. Algebra 13(1969): 119-136.

# 5
# POLYNOMIAL IDENTITIES

If $R$ is a simple ring with involution in which any two symmetric elements commute we saw, as a consequence of Theorem 2.1.6, that $R$ is at most 4-dimensional over its center. Put another way, if $S$ satisfies the polynomial identity $p(x, y) = xy - yx$ then $R$ must satisfy the standard identity $s_4(x_1, \ldots, x_4)$. We saw another phenomenon of this type in Corollary 2 to Theorem 2.1.11. There, if $K$ is finite dimensional over $Z^+ = Z \cap S$, of dimension $n$, say, then, if $R$ is simple, $R$ must be of dimension at most $n + n^2$ over $Z$. In this situation $K$, of course, satisfies the standard identity $s_{n+1}$ and $R$ ends up finite dimensional, and hence satisfies a polynomial identity.

Such considerations led us to conjecture, about 1958, that if $R$ is a simple ring in which $S$ (or $K$) satisfies a polynomial identity then $R$ must be finite dimensional over its center, this dimension being bounded by a simple function of the degree of the polynomial identity satisfied by $S$. In 1966 we proved this, the result appearing in [3], when the characteristic of $R$ is not 2.

Martindale [4] picked the problem up from here and proved that any semi-prime algebra with involution of characteristic not 2 whose symmetric elements satisfy a polynomial identity of degree $d$ must satisfy the standard identity of degree $2d$. Amitsur [1], using the technique which we have called the "Amitsur trick" in our proof of Theorem 1.6.1 pushed the result through to its final form. Moreover, in his proof, characteristic 2 did not have to be excluded. The result he proved is: if R is an algebra with involution over a commutative ring $\Omega$ in which the symmetric (or skew) elements satisfy a polynomial identity of degree d then R satisfies $s_{2d}^m$ for some m, where $s_{2d}$ is the standard identity of degree 2d.

This result, aside from its intrinsic beauty, is highly useful and has been applied often. However, Amitsur was not satisfied with it and felt that the direction of the theorem was not the proper one, that the result should be a special case of a more general and global phenomenon. In this he was certainly correct, as he showed in [2], where he proved what seems to us to be the right theorem for this kind of setting. His result is: if R is an algebra with involution over a commutative ring $\Omega$ and R satisfies an identity of the form $p(x_1, \ldots, x_r; x_1^*, \ldots, x_r^*)$ of degree d, then R satisfies $s_{2d}(x)^m$ for some m. Of course the results we stated above are all special cases of this theorem.

This chapter is devoted to the proof of this striking theorem of Amitsur. The proof we give here is due to Susan Montgomery. Although it is much in the spirit of Amitsur's proof, and some of the early steps are identical to his, we feel that this proof is easier and more transparent.

In what follows, R will be a ring with involution $*$ which is an algebra over a commutative ring C. Let C[x, y] be the free algebra over C in the infinite set of non-commuting variables $x_i, y_i$. In C[x, y] we can generate an involution by $x_j^* = y_j$ and $y_j^* = x_j$ and

and linearity and by insisting that $*$ be an anti-automorphism. Thus, if $p(x, y) = p(x, x^*) \in C[x, y] = C[x, x^*]$ we know what is meant by the degree of $p(x, y) = p(x, x^*)$ and by the statement "R satisfies $p(x, x^*)$."

We assume henceforth that R satisfies $p(x, x^*)$ of degree d, that is, $p(a, a^*) = 0$ for $a \in R^n$. To avoid niceties about non-triviality of the polynomial, we assume that some monomial of degree d occurs in $p(x, x^*)$ with coefficient 1.

As usual in most work with polynomial identities, our first step is to show that R satisfies a homogeneous multilinear identity of a particular form. This is

LEMMA 5.1.1. If R satisfies $p(x, x^*) = \alpha m(x, x^*) + \ldots$ of degree d, where $\alpha \neq 0 \in C$ and $m(x, x^*)$ is a monomial of degree d, then R satisfies $p(x, x^*) = \alpha x_1 x_2 \cdots x_d + q(x, x^*)$ where each monomial of $q(x, x^*)$ is of degree d, involves each $x_i$ or $x_i^*$ but not both, and where $x_1 x_2 \cdots x_d$ does not occur in $q(x, x^*)$.

<u>Proof.</u> Let $t_i(p_1(x, x^*)$ be the sum of the degrees of $p(x, x^*)$ in $x_i$ and in $x_i^*$. We proceed by induction on $t_i(p_1)$. If $t_i(p_1) = k$, and if $h(x, z; x^*, z^*)$ is the linearization of $p_1$ on $x_i$ defined by

$$h(x, z; x^*, z^*) = p_1(x_1, \ldots, x_i + z, \ldots, x_r; x_1^*, \ldots, x_i^* + z^*, \ldots, x_r^*)$$
$$-p_1(x_1, \ldots, x_i, \ldots, x_r; x_1^*, \ldots, x_i^*, \ldots, x_r^*)$$
$$- p(x_1, \ldots, z, \ldots, x_r; x_1^*, \ldots, z^*, \ldots, x_r^*)$$

then $t_i(h) \leq k-1$ and there is a monomial of degree d in h of the form $\alpha m'$. By induction we get the result, since R also satisfies $h(x, z; x^*, z^*)$. We thus get a multilinear identity in $x_1, \ldots, x_d$ and $x_1^*, \ldots, x_d^*$. By the linearity we have that each monomial can contain at at most one of each $x_i$ or $x_i^*$. If we pick p of minimal degree we also get, per usual, that p is homogeneous. Renumbering the variables — if we need to, replace some $x_i$ by $x_i^*$ — we get that R satisfies a polynomial identity of the form described in the assertion of the lemma.

Since we are assuming that some monomial of degree d in $p(x, x^*)$ occurs with coefficient 1, by the lemma we may assume that R satisfies a polynomial identity of the form

(1) $$p(x, x^*) = x_1 x_2 \cdots x_d + q(x, x^*)$$

where each monomial in $q(x, x^*)$ is of degree d and involves each $x_i$ or $x_i^*$ but not both, and where $x_1 x_2 \cdots x_d$ does not occur in $q(x, x^*)$

Henceforth, when we assume that R satisfies a polynomial identity of the type $p(x, x^*)$, we shall assume that $p(x, x^*)$ is of the form described in (1).

As an immediate consequence of this multilinearity of $p(x, x^*)$ we have the

COROLLARY. If R is as in Lemma 5.1.1 and if A is a commutative C-algebra then $R \otimes_C A$, relative to the involution defined by $(r \otimes \alpha)^* = r^* \otimes \alpha$, satisfies $p(x, x^*)$. In particular, if $A = C[x]$ is the polynomial ring over C, then $R \otimes_C A$ satisfies $p(x, x^*)$.

With these usual formalities out of the way we can get down to the essentials of the proof. Our first result holds for any primitive ring with involution.

LEMMA 5.1.2. Let $R$ be a primitive ring with involution and suppose that $R$ acts faithfully and densely on $V$, which is a left vector space over a division ring $D$. Then, either

1. $R$ has a minimal left ideal

or

2. If $W \subset V$ is a subspace finite dimensional over $D$ and if $v \notin W$, there exists an $x \in R$ such that $xW = x^*W = 0$, $x^*v = 0$ but $xv \notin W + Dv$.

Proof. Suppose that $R$ does not have a minimal left ideal; then $R$ has no non-zero elements of finite rank on $V$. That is, $rV$ is infinite dimensional over $D$ for all $r \neq 0$ in $R$.

Let $W$ be a finite dimensional subspace of $V$, and suppose that $v \notin W$ where $v \in V$. Since $R$ acts densely on $V$ there exists $a \neq 0$ in $R$ such that $aW = 0$ and $av = 0$. Since $a \neq 0$, $a^*V \neq 0$ by the faithful action of $R$ on $V$, hence $a^*V$ cannot be finite dimensional over $D$. Thus there is a $u \in a^*V$ with $u \notin W + Dv$.

Let $B = \{r \in R \mid rW = 0\}$; $B$ is a left ideal of $R$ and, since $v \notin W$, $Bv \neq 0$ by the density of $R$ on $V$. Therefore $V = Bv$. Hence $u \in a^*V = a^*Bv$, and so $u = a^*bv$ for some $b \in B$. Let $x = a^*b$; $x \in B$ because $B$ is a left ideal of $R$ and $xW = a^*bW = 0$. Also, $x^* = b^*a$, thus $x^*v = b^*av = 0$ since $av = 0$. Because $xv = u \notin W + Dv$ we have produced the $x$ of the assertion of the lemma.

We now come to the first important step in the proof, which interrelates the existence of an identity of type $p(x, x^*)$ with the structure of the ring.

LEMMA 5.1.3. Let $R$ be a primitive ring with involution and suppose that $R$ satisfies $p(x, x^*)$ of degree $d$. Then $R$ has a minimal left ideal.

<u>Proof.</u> As we pointed out, we may assume that $p(x, x^*) = x_1x_2 \cdots x_d + q(x, x^*)$ of the form described in (1). Since $R$ is primitive it acts faithfully and densely on a left vector space $V$ over a division ring $D$. If Lemma 5.1.3 is not valid, by virtue of Lemma 5.1.2, with $W = 0$ and $v \neq 0 \in V$ we can find an element $t_1 \in R$ such that $t_1^* v = 0$ but $t_1 v \notin Dv$.

We are going to construct, inductively, a sequence $t_1, \ldots, t_n \ldots$ in $R$ such that:

a) $t_j(t_i \cdots t_2t_1)v = 0$ for $j > i+1$

b) $t_j^*(t_i \cdots t_2t_1)v = 0$ for $j \geq i+1$

c) the set $v, t_1v, t_2t_1v, \ldots, t_n \cdots t_2t_1v$ is linearly independent over $D$.

Suppose we have such a set for $n = k$, $t_1, \ldots, t_k$. Let $U = Dv + Dt_1v + \ldots + Dt_{k-1} \cdots t_2t_1v$ and let $v_k = t_kt_{k-1} \cdots t_1v$; by our induction hypothesis, $v_k \notin U$. Hence, by Lemma 5.1.2, there is an element $t_{k+1} \in R$ such that $t_{k+1}U = t_{k+1}^*U = 0$, $t_{k+1}^*v_k = 0$ but $t_{k+1}v \notin U + Dv_k$. Therefore $t_1, \ldots, t_{k+1}$ satisfy (a), (b) and (c) and the inductive construction is complete for any $n$. We use $n = d$, the degree of $p(x, x^*)$.

Let $x_1 = t_d$, $x_2 = t_{d-1}, \ldots, x_d = t_1$ and substitute these values in

$$0 = p(x, x^*)v = t_dt_{d-1} \cdots t_1v + q(t_d, \ldots, t_1; t_d^*, \ldots, t_1^*)v .$$

We shall show that $q(t_d, \ldots, t_1; t_d^*, \ldots, t_1^*)v = 0$ and thus get the contradiction $t_d t_{d-1} \cdots t_2 t_1 v = 0$. Any monomial in $q(t_d, \ldots, t_1; t_d^*, \ldots, t_1^*)$ not ending in $t_1$ annihilates $v$ by our construction of the sequence $t_1, \ldots, t_d$. Thus the only monomials in $q(t_d, \ldots, t_1; t_d^*, \ldots, t_1^*)$ which give a non-zero contribution to $q(t_d, \ldots, t_1; t_d^*, \ldots, t_1^*)v$ must end in $t_1$. By the nature of $q(x, x^*)$ such a monomial ending in $t_1$ cannot contain $t_1^*$, thus only $t_d, \ldots, t_2$, $t_d, \ldots, t_2, t_d^*, \ldots, t_2^*$ appear. Hence

$$q(t_d, \ldots, t_1; t_d^*, \ldots, t_1^*)v = q_1(t_d, \ldots, t_2; t_d^*, \ldots, t_2^*)t_1 v \, ,$$

where the monomials in $q_1$ involve each $x_i$ or $x_i^*$ but not both, for $2 \leq i \leq d$. Repeat the reasoning on $q_1(t_d, \ldots, t_2; t_d^*, \ldots, t_2^*)t_1 v$ to get it of the form $q_2(t_d, \ldots, t_3; t_d^*, \ldots, t_3^*)t_2 t_1 v$. Continuing, since $x_1 x_2 \cdots x_d$ is not present in $q(x, x^*)$ we eventually get that $q(t_d, \ldots, t_1; t_d^*, \ldots, t_1^*)v = 0$, and so $t_d t_{d-1} \cdots t_1 v = 0$, in contradiction to the nature of the $t$'s. This proves the lemma.

We are able, by the preceding lemma, to intertwine the presence of the identity $p(x, x^*)$ on $R$ with the nature of the primitive images of $R$. This is the key step in the proof.

LEMMA 5.1.4. Let $R$ be a ring with involution satisfying $p(x,x^*)$ of degree $d$. If $P$ is a primitive ideal of $R$ then $R/P$ satisfies a polynomial identity of degree 2d, hence is a simple algebra of dimension at most $d^2$ over its center.

<u>Proof.</u> The argument divides according as $P^* = P$ or $P^* \neq P$. The former is the easy part, the latter requires considerably more work.

Suppose, first, that $P^* \not\subset P$. Thus $(P^* + P)/P$ is a non-zero ideal in $R/P$; every element $u_i \in P^*$ has $u_i^* \in P$ hence, since $p(x, x^*) = x_1 \cdots x_d + q(x, x^*)$, using $u_1, \ldots, u_d \in P^*$ we have, in $R/P$, that $0 = \overline{u}_1 \cdots \overline{u}_d + q_1(\overline{u}_1, \ldots, \overline{u}_d)$ where $q_1$ is that part of $q$ which does not involve any of the $x_i^*$. Hence $A = (P^* + P)/P$ satisfies $x_1 x_2 \ldots x_d + q_1(x_1, \ldots, x_d)$ of degree $d$. As a non-zero ideal in a primitive ring, $A$ itself is primitive. Because it satisfies a P.I., by Kaplansky's theorem, $A$ is finite dimensional simple, hence has a unit element $f$. But then $f$ is a central idempotent in $R/P$, thus is the unit element of $R/P$. In consequence, $A = R/P$, and so $R/P$ satisfies a polynomial identity of degree $d$.

We thus may suppose that $P^* = P$. Therefore $R/P$ is a primitive ring, and the involution of $R$ induces an involution on $R/P$, such that $R/P$ satisfies $p(x, x^*)$. Hence, to finish the proof, we must show that a primitive ring with involution satisfying $p(x, x^*)$ of degree $d$ must satisfy a polynomial identity of degree $2d$. For the rest of the proof we assume that $R$ is primitive, therefore.

By Lemma 5.1.3, $R$ has a minimal left ideal. By Theorem 1.2.2, $R$ acts faithfully and densely on a vector space over a division ring $D$, where $V$ has an inner product which is either Hermitian or alternate, and $R$ contains all continuous transformations, re this inner product, of finite rank.

Our first objective is to show that $V$ is at most $d$-dimensional over $D$. We divide the argument according as the inner product is Hermitian or alternate.

If the inner product is Hermitian, by Theorem 1.2.2, $R \supset D_m$, the $m \times m$ matrices over D, with $*$ inducing an involution on $D_m$ of transpose type, for all $m \leq \dim_D V$. Suppose that $d < m$, that is, that $\dim_D V > d$. We can pick matrix units $e_{ij}$ for $D_m$ in this case such that $e_{ij}^* = \alpha_{ij} e_{ji}$ where $\alpha_{ij} \in D$. Let $t_i = e_{i,i+1}$ for $i = 1, 2, \ldots, d$; thus

$$0 = p(t_1, \ldots, t_d; t_1^*, \ldots, t_d^*) = t_1 \cdots t_d + q(t_1, \ldots, t_d; t_1^*, \ldots, t_d^*)$$

$$= e_{1,d+1} + q(t_1, \ldots, t_d; t_1^*, \ldots, t_d^*).$$

If some $t_i$ and some $t_j^*$ occur in the same monomial of $q$, by the nature of $q$, $i \neq j$. But then $t_i t_j^* = \alpha_{j,j+1} e_{i,i+1} e_{j+1,j} = 0$ hence the contribution of such a monomial to $q(t, t^*)$ is 0. Therefore the only monomials giving a non-zero contribution to $q(t, t^*)$ can involve only $t_i$'s or only $t_i^*$'s. From the nature of the $t_i$'s, we get a non-zero contribution only if the $t_i$'s are in ascending order, or the $t_i^*$'s in descending order. Because $x_1 \cdots x_d$ does not occur in $q(x, x^*)$, the only term in $q(x, x^*)$ giving a non-zero contribution to $q(t, t^*)$ is $\gamma x_d^* \cdots x_1^*$, and this gives $\gamma t_d^* \cdots t_1^* = \gamma_1 e_{d+1,1}$ where $\gamma_1 \in D$. But this gives the contradiction $0 = e_{1,d+1} + \gamma_1 e_{d+1,1}$. We therefore must have that $\dim_D V \leq d$; this finishes the Hermitian case.

We may thus suppose that the form is alternate on V; in this case D must be a field F and, for every m with $2m \leq \dim_F V$, $R \supset F_{2m}$ the $2m \times 2m$ matrices over F. Moreover, the $*$ of R induces the symplectic involution on $F_{2m}$, given by $(A_{ij})^* = (A_{ji}^*)$ where $A_{ij}$ are $2 \times 2$ matrices over F, and $\begin{pmatrix} \alpha & \beta \\ \gamma & \delta \end{pmatrix}^* = \begin{pmatrix} \delta & -\beta \\ -\gamma & \alpha \end{pmatrix}$ for these $2 \times 2$ matrices.

We claim that $2m \leq d$. Consider, first, the case when $d$ is even, and suppose that $d < 2m$. We pick $t_1, \ldots, t_d$ as follows:

$$t_i = e_{ii} \qquad \text{if } i \text{ is odd,}$$

$$t_i = e_{i-1,i+1} \qquad \text{if } i \text{ is even.}$$

By the choice made, all the subscripts of the $t_i$ are odd. Computing the $*$ of these elements we have:

$$t_i^* = e_{i+1,i+1} \qquad \text{if } i \text{ is odd,}$$

$$t_i^* = e_{i+2,i} \qquad \text{if } i \text{ is even.}$$

The subscripts of the $t_i^*$'s are all even.

Substituting these values in $p(x, x^*)$ we get $0$ if a monomial contains an $x_i$ and an $x_j^*$, since one gives odd subscripts and the other even. Hence, as before, the contributions can come only from $x_1x_2\cdots x_d$ and $x_d^*\cdots x_2^*x_1^*$. Thus

$$0 = p(t_1, \ldots, t_d; t_1^*, \ldots, t_d^*) = t_1t_2\cdots t_d + t_d^*\cdots t_2^*t_1^*$$

$$= e_{11}e_{13}e_{33}\cdots e_{d-1,d-1} + \gamma e_{dd}\cdots e_{44}e_{42}e_{22}$$

$$= e_{1,d-1} + \gamma e_{d,2} \neq 0 ,$$

a contradiction. A similar argument works when $d$ is odd. Thus $2m \leq d$, and so $\dim_D V \leq d$.

In the alternate case we have that $R \approx F_n$ where $n \leq d$. In the Hermitian case we want something similar. We already know that $R \approx D_n$ where $n \leq d$. If $*$ is of the second kind, then every element in $R$ is a central combination of symmetric elements; then easily, using this fact, we get that $R$ satisfies $p(x_1, \ldots, x_d; y_1, \ldots, y_d)$ of degree $2d$. Suppose then that $R$ is of the first kind. Let $F$ be a maximal subfield of D and consider $R\otimes_Z F$; then R is primitive, by the corollary to Lemma 6.5.1 satisfies $p(x,x^*)$, and the commuting ring of $R\otimes_Z F$ is F. By the argument above, $R\otimes_Z F \approx F_n$ for some $n \leq d$, thus it satisfies an identity of degree 2d by the Amitsur-Levitzki theorem, hence R also does. The lemma is now proved.

COROLLARY 1. If R is a semi-simple ring with involution and satisfies $p(x,x^*)$ of degree d then R satisfies the standard identity of degree 2d.

If R has no nil ideals and satisfies $p(x,x^*)$ of degree d then $R[\lambda]$, the polynomial ring in $\lambda$ over R is semi-simple, and by the corollary to Lemma 6.5.1, satisfies $p(x,x^*)$.

By the corollary above, $R[\lambda]$ satisfies $s_{2d}(x_1,\ldots,x_{2d})$, hence R must also. Therefore

COROLLARY 2. If R is without nil ideals and satisfies $p(x,x^*)$ of degree d then R satisfies the standard identity of degree 2d.

Suppose that R is semi-prime with involution and satisfies $p(x,x^*)$ of degree d. We claim that R has no nil ideals. For, suppose that $I \neq 0$ is a nil ideal; we may suppose that $I = I^*$ without loss of generality, hence we may assume that there is an element $a \neq 0 \in I$ with $a^2 = 0$ where $a^* = a$ or $a^* = -a$. We may suppose that $p(x,x^*)$ is of the form (1). Let $r_1,\ldots,r_d \in R$ and let $t_i = ar_i$, thus $t_i^* = r_i^* a^* = \pm r_i^* a$. From $0 = p(ar_1,\ldots,ar_d; r_1^*a,\ldots,r_d^*a)$ we have $p(ar_1,\ldots,ar_d; r_1^*a,\ldots,r_d^*a)aR = 0$. But in $p(ar_1,\ldots,ar_d; r_1^*a,\ldots,r_d^*a)a = 0$ we get a zero contribution from any term involving an $x_i^*$. Thus $p(ar_1,\ldots,ar_d; r_1^*a,\ldots,r_d^*a)aR$ reduces to

$p_1(ar_1, \ldots, ar_d)aR = 0$ for some polynomial $p_1(y_1, \ldots, y_d) \neq 0$. Thus $aR \neq 0$ is a nil left ideal of R satisfying a polynomial identity. By Lemma 2.1.1, R must have a nilpotent ideal, contrary to the semi-primeness of R. Thus we have proved

LEMMA 5.1.5. If R is semi-prime and satisfies $p(x, x^*)$ of degree d then R satisfies the standard identity of degree 2d.

We have all the pieces to prove the lovely theorem due to Amitsur [2]

THEOREM 6.5.1. If R satisfies $p(x, x^*)$ of degree d then R satisfies $s_{2d}(x)^m$ for some m, where $s_{2d}(x)$ is the standard identity in 2d variables.

<u>Proof.</u> To pass from the semi-prime case to the general case use the Amitsur trick exactly as was done in the proof of Theorem 1.6.1. The argument goes over <u>in toto</u> and proves the theorem.

Let's consider a particularly easy case, namely a ring R which is normal in the sense that $xx^* = x^*x$ for all $x \in R$. By the theorem, R satisfies $s_4(x_1, \ldots, x_4)^m$ and if R is semi-prime then R satisfies $s_4(x_1, \ldots, x_4)$, so is imbeddable in $2 \times 2$ matrices over a commutative ring.

This result, with $xx^* = x^*x$ for all $x \in R$, is obtainable also by completely elementary, formal arguments.

We now specialize Theorem 6.5.1 to the special cases which were the origins of Theorem 6.5.1.

THEOREM 6.5.2. Let R be a ring with involution such that

1. S satisfies $p(x_1, \ldots, x_r)$ of degree d.

or

2. K satisfies $p(x_1, \ldots, x_r)$ of degree d.

Then R satisfies $s_{2d}(x)^m$ for some m where $s_{2d}(x)$ is the standard identity in 2d variables. If R is semi-prime, $m = 1$.

Proof. If S satisfies $p(x_1, \ldots, x_r)$ then R satisfies $q(x, x^*) = p(x_1 + x_1^*, \ldots, x_r + x_r^*)$ of degree d; if K satisfies $p(x_1, \ldots, x_r)$ then R satisfies $q(x, x^*) = p(x_1 - x_1^*, \ldots, x_r - x_r^*)$ of degree d. Applying Theorem 6.5.1, the results drop out.

There are results for generalized polynomial identities for rings with involution; we refer the reader to Martindale's paper [5].

Bibliography

1. S.A. Amitsur. Rings with involution. Israel Jour. Math. 6 (1968): 99-106.

2. S.A. Amitsur. Identities in rings with involution. Israel Jour. Math. 7 (1969): 63-68.

3. I.N. Herstein. Special simple rings with involution. Jour. Alg. 6 (1967): 369-375.

4. W. Martindale. Rings with involution and polynomial identities. Jour. Alg. 11 (1969): 186-194.

5. W. Martindale. Prime rings with involution and generalized polynomial identities. Jour. Alg. 22 (1972): 502-516.

# 6
# POTPOURRI

In this closing chapter we shall present some recent results which, while they do have a certain flavor in common, are, by and large, not related to each other. For most of these results the unifying theme is that of dichotomy, a dichotomy of the type expressed in the Brauer-Cartan-Hua theorem. Our objective will be, in most of these situations, to show that certain subgroups or subrings of a ring with involution which are invariant with respect to certain natural operations must be small or large, in a very well specified way. By small we shall mean "contained in the center", by large we shall mean "contains a non-zero ideal of the ring".

## 1. A Unitary Version of the Brauer-Cartan-Hua Theorem

The Brauer-Cartan-Hua theorem states that the only subdivision rings of a division ring which are invariant with respect to all the inner automorphisms are the division ring itself and subfields of the center of the division ring. In this section we shall prove an analog of this result for division rings with involution.

In any ring with involution which has a unit element, we call an element $u$ unitary if $uu^* = u^*u = 1$. What we shall be concerned with here will be invariance with respect to conjugation by the unitaries. Before we can even get started on such an investigation, we need a ready source ofunitaries. In our context this is provided us by the elements $u = (1-k)(1+k)^{-1}$, where $k^* = -k$ is in $K$, which are unitary. This is just the Cayley parametrization. If char $D \neq 2$ we obtain lots of unitaries this way; in fact we get enough such elements to push the theorem through.

If char $D = 2$ the whole matter is up in the air at present. It seems likely that Theorem 6.1.1 is false in that case. In fact a proposed counter-example was suggested in characteristic 2; its nature was such that the only unitary element in $D$ was 1, in which case invariance with respect to the unitaries becomes a vacuous condition. But there are doubts about the counter-example at present. However we feel that it should be possible to construct a division ring with involution, of characteristic 2, in which 1 is the only unitary element.

It would be interesting and useful to generalize our result here to simple rings with involution. The kind of condition that one might impose could be that all $1 + k$ are invertible for $k \in K$, or that the ring has an idempotent $e \neq 0, 1$ such that $e^* = e$. For certain rings of operators, Topping did this [17]. However, it would be useful to have such results in the fairly general context of simple rings with involution subject to some mild conditions.

The result which we shall prove here comes from [5]. However, we shall not prove the theorem in its entirety. Characteristic 3 introduces some nasty complications in the proof, and makes the argument a good deal more intricate. In our proof here we shall assume that the characteristic is not 3, in order to simplify things, although we shall later use the result in its full generality. The reader who wishes to see the details of the proof in general is advised to look at the original source [5].

We begin with a very special case of theorems that shall be proved later in this chapter. However, this special case will do in the proof of the main theorem of this section.

LEMMA 6.1.1. Let $D$ be a division ring with involution such that $\operatorname{char} D \neq 2, 3$. Suppose that $a \in D$ commutes with $ak - ka$ for all $k \in K$. Then, if $\dim_Z D > 4$, $a$ must be in $Z$, the center of $D$.

<u>Proof.</u> Define $\delta: D \to D$ by $\delta(x) = ax - xa$; $\delta$ is a derivation on $D$. Our hypothesis on $a$ translates into $\delta^2(k) = 0$ for all $k \in K$. But, if $k \in K$, then $k^3 \in K$; hence $\delta^2(k^3) = 0$. Expanding this last relation by making use of Leibniz's rule and $\delta^2(k) = 0$ we obtain $2\delta(k)^2 k + 2\delta(k)k\delta(k) + 2k\delta(k)^2 = 0$. Apply $\delta$ to this; using $\delta^2(k) = 0$ we end up with $6\delta(k)^3 = 0$. Since $D$ is a division ring and $\operatorname{char} D \neq 2,3$ we obtain $\delta(k) = 0$, which is to say, $ak = ka$ for all $k \in K$. But, since $\dim_Z D > 4$, by Theorem 2.1.10, $K$ generates $D$, in consequence of which, $a \in Z$ follows.

We can now state and prove the unitary version of the Brauer-Cartan-Hua theorem.

THEOREM 6.1.1. Let $D$ be a division ring with involution, of characteristic different from 2. Let $A$ be a subdivision ring of $D$ such that $uAu^{-1} \subset A$ for every unitary element $u$ in $D$. Then

1. if $A$ is commutative and $\dim_Z D > 4$, $A \subset Z$.

2. if $A$ is not commutative and $\dim_Z D > 16$, $A = D$.

Proof. We remind the reader that in our proof we shall assume that $\text{char } D \neq 3$; the theorem is true even if $\text{char } D = 3$.

Let $k \in K$; then $u = (1-k)(1+k)^{-1}$ is unitary. Hence, if $a \in A$, we have that $a - uau^{-1} \in A$, that is, $a - (1-k)(1+k)^{-1}a(1+k)(1-k)^{-1}$ is in $A$. A simple calculation reveals that this last element is $2(1+k)^{-1}(ka - ak)(1-k)^{-1}$, the net result of which is that $b = (1+k)^{-1}(ka - ak)(1-k)^{-1}$ is in $A$ for $a \in A$, $k \in K$.

Our first objective is to show that $ka - ak \in A$ for all $a \in A, k \in K$. If $ka = ak$ this is certainly true. On the other hand, if $ka - ak \neq 0$ then, since $A$ is a subdivision ring of $D$, $b^{-1} = (1-k)(ka - ak)^{-1}(1+k) \in A$. If $c = (ka - ak)^{-1}$ this says that

(1) $\qquad c - (kc - ck) - kck \in A.$

If, instead of using $k$ we use $-k$ in the argument above (since $c$ becomes replaced by $-c$) we obtain

(2) $\qquad c + (kc - ck) - kck \in A.$

From (1) and (2) we get that

(3) $\qquad c - kck \in A.$

In (3) replace $k$ by $2k$; $c$ is then replaced by $\frac{1}{2}c$. We thus get that $\frac{1}{2}c - (2k)(\frac{1}{2}c)(2k) \in A$, that is ,

(4) $\qquad c - 4kck \in A.$

From (3) and (4) we obtain that $3c \in A$, and, since $\operatorname{char} D \neq 3$ (as we are assuming), we get that $c \in A$. Hence $c^{-1} = ka - ak$ is in $A$. In short, we have attained our first objective, namely, that $[A, K] \subset A$.

If $A$ is commutative, using Lemma 6.1.1, we obtain that $A \subset Z$ provided that $\dim_Z D > 4$. This finishes the proof of part 1 of the theorem.

We now turn to part 2. Suppose that $\dim_Z D > 16$ and that $A \neq D$. Suppose further, for the moment, that $A = A^*$. If every element in $A$ is symmetric then $A$ is commutative, so by the above, $A \subset Z$. So, if $A \not\subset Z$, there is an $a \in A$ such that $a - a^* \neq 0$. Hence $A^- = A \cap K \neq 0$. Since $[A, K] \subset A$ we have that $[A^-, K] \subset A^-$, which is to say, $A^-$ is a Lie ideal of $K$. Since $\dim_Z D > 16$, we have by Theorem 2.12 of TRT that either $A^- \subset Z$ or $A^- \supset [K, K]$.

If $A^- \subset Z$, and $\alpha \neq 0$ is in $A^-$ then, for $a \in A$, $\alpha(a + a^*) \in A^- \subset Z$ giving us $a + a^* \in Z$. Together with $a - a^* \in Z$ we get $a \in Z$, and so $A \subset Z$.

On the other hand, if $A \supset [K, K]$, then, since by Theorem 2.13 of TRT, $[K, K]$ generates $D$, we get $A = D$, contrary to $A \neq D$. Thus, if $A = A^*$ the theorem is proved.

We now do the general case. Suppose that $\dim_Z D > 16$ and $A \neq D$. Thus $A \cap A^* \neq D$ is invariant re $*$, and invariant re the unitaries. By the above, $A \cap A^* \subset Z$. Also, from the earlier part of the proof, we know that $[A, K] \subset A$, $[A^*, K] \subset A^*$.

Let $a, b \in A$; then $a(b - b^*) - (b - b^*)a \in A$, hence $ab^* - b^*a \in A$. From $(a - a^*)b^* - b^*(a - a^*) \in A^*$ we get $ab^* - b^*a \in A^*$. Hence $\mu = ab^* - b^*a$ is in $A \cap A^* \subset Z$. Also, $a^2b^* - b^*a^2 \in Z$; but

$a^2b^* - b^*a^2 = 2\mu a$, hence, if $\mu \neq 0$, we have $a \in Z$. But in that case $\mu = ab - ba = 0$. In other words, $\mu = 0$, which is to say, $ab^* = b^*a$ for all $a, b \in A$.

Let $k = ab^* - ba^*$; since $k \in K$, $A \ni ak - ka = -(ab - ba)a^*$ since $ab^* = b^*a$. If $ab - ba \neq 0$ we then get $a^* \in A$ and so $a^* \in A \cap A^* \subset Z$, whence $a \in Z$. But then $ab = ba$. In other terms, we have shown that if $A \neq D$ then $A$ must be commutative. By what we did previously, we have that $A \subset Z$. The theorem is proved.

Some remarks are in order. First, if $\dim_Z D = 4$ we can give an example of a non-central, commutative subdivision ring invariant re all the unitaries. Let $D$ be the quaternions over the real field and let $*$ be defined by $(\alpha_0 + \alpha_1 i + \alpha_2 j + \alpha_3 k)^* = \alpha_0 - \alpha_1 i + \alpha_2 j + \alpha_3 k$. The subdivision ring $A = \{\alpha_0 + \alpha_1 i\}$ is easily shown to be invariant re conjugation by the unitaries.

Second, if $\dim_Z D = 16$, we can give an example of a subdivision ring $A$, $A \not\subset Z$, $A \neq D$, such that $uAu^{-1} \subset A$ for all unitaries $u$. Over a suitable field $F$ (a rational function field in 4 variables will do) we can construct two 4-dimensional division algebras, $D_1$ and $D_2$, such that $D = D_1 \otimes_F D_2$ is a division algebra over $F$. It is now easy to introduce an involution on $D$ such that $D_1$ is invariant re all unitaries.

To verify these last comments it is enough to produce a subdivision ring $A$, neither in $Z$ nor equal to $D$, such that $[A, K] \subset A$. For, in that case, if $a \in A$, $k \in K$ and $ka - ak \neq 0$, let $b = (ak - ka)^{-1}$. Then $kbk = k(ak - ka)^{-1}k = (k^{-1}a - ak^{-1})^{-1}$, and since $k^{-1} \in K$, $k^{-1}a - ak^{-1} \in K$ from which we have $(k^{-1}a - ak^{-1})^{-1} \in A$. Thus

$$c = (1+k)(ak - ka)^{-1}(1 - k) = (ak - ka)^{-1} + k(ak - ka)^{-1} - (ak - ka)^{-1}k - k(ak - ka)^{-1}k$$

is in A, and so $c^{-1} = (1-k)^{-1}(ak - ka)(1+k) \in A$. But,

$$2(ak - ka) = ((1-k)a(1+k) - (1+k)a(1-k))$$

whence

$$(1-k)^{-1}((1-k)a(1+k) - (1+k)a(1-k)(1+k)^{-1} \in A,$$

giving us

$$(1-k)^{-1}(1+k)a(1+k)^{-1}(1-k) \in A .$$

However, the general unitary element $u \neq -1$ can be written as $u = (1-k)^{-1}(1+k)$ for some $k \in K$. This gives us that A is invariant with respect to conjugation by the unitaries.

If D has an involution of the second kind, Theorem 6.1.1 can be slightly strengthened. Here, if A is a commutative subdivision ring invariant with respect to conjugation by the unitaries then A must be in the center of D. Also, if $\dim_Z D > 4$ and A is invariant re the unitaries but A is not commutative, then $A = D$.

Similar statements can be made, for involutions of the second kind, even if char $D = 2$, provided Z is <u>not</u> the field of four elements.

Finally, if $\dim_Z D = 4$ and $*$ is of the second kind, it is easy to produce examples of subdivision rings A invariant with respect to conjugation by the unitaries with $A \not\subset Z$ and $A \neq D$.

## 2. *-Radicality in Division Rings

Earlier, in Theorems 3.2.1, 3.2.2, we discussed the structure of division rings with involution where a power of every symmetric element (or, a power of every skew element) falls in the center. These results were the analogs, for division rings with involution, of a well-known result of Kaplansky [11] which states that a division ring is commutative if a power of every element falls in the center.

There is an extension of Kaplansky's theorem, due to Faith [4], which concludes that a division ring is commutative if a power of every element falls in a given, proper subring. Loustau [15] gave a partial extension of this for division rings with involution, proving the following rather special theorem: Let **D** be a division ring with involution, char $D \neq 2$, and $A$ a commutative subring of $D$. If for every $s \in S$, $s^{n(s)} \in A$, where $n(s) \geq 1$, then $D$ is 4-dimensional over its center and $S \subset Z$.

We propose to obtain the direct analog of Faith's result for division rings with involution; Loustau's theorem will emerge as a very special instance of the theorem we obtain. We also derive, from this result, a fairly general "commutativity-type" theorem for division rings with involution. En route to these results we need to develop the analog of the hypercenter (see [6]) for symmetric and skew elements in division rings. Also these results come from a joint work by Chacron and ourselves [2].

$D$ will be a division ring with involution, $Z$ its center, $S, K$, as usual, the symmetric and skew elements respectively, and, finally, $N = \{xx^* \mid x \in D\}$ will denote the set of norms in $D$. Recall that if char $D \neq 2$ the statement $N \subset Z$ is synonomous with the statement $S \subset Z$.

We begin with

LEMMA 6.2.1. a) If in D, given $a, b \in S$ there exists an integer $n = n(a, b) \geq 1$ such that $ab^n = b^n a$, then $N \subset Z$ and $\dim_Z D \leq 4$.

b) If in D, given $a, b \in K$, $ab^n = b^n a$ for some integer $n = n(a, b) \geq 1$, then $k^2 \in Z$ for all $k \in K$, and $\dim_Z D \leq 4$.

Proof. We only prove Part a) in detail; the proof of Part b) proceeds exactly in the same manner, invoking Theorem 3.2.2 rather than Theorem 3.2.1.

Suppose that $a, b \in S$; let $D_1$ be the subdivision ring generated by a and b. Clearly $D_1^* = D_1$. If $t = t^*$ is in $D_1$ then, since $t^m a = at^m$ and $t^n b = bt^n$ for some $m, n \geq 1$, $t^{mn}$ commutes with both a and b; since a and b generate $D_1$, we have that $t^{mn} \in Z(D_1)$. By Theorem 3.2.1 we know that all norms in $D_1$ must then be in $Z(D_1)$; since $b^2$ is a norm in $D_1$, we have that $ab^2 = b^2 a$. If $S \subset Z$ then, since $N \subset S \subset Z$, we certainly have that $N \subset Z$. On the other hand, if $S \not\subset Z$ then S generates D; by the above we have that $b^2 \in Z$ for all $b \in S$. By Theorem 3.2.1 (or directly) we have that $N \subset Z$. This proves the lemma.

We now make the

DEFINITION. We say that D is S-radical over the subring $A \neq D$ if, for every $s \in S$, $s^{n(s)} \in A$ for some $n(s) \geq 1$.

If D is S-radical over A and over B it is trivial that D is S-radical over $A \cap B$. Hence, if D is S-radical over A, it is also

S-radical over $A^*$, and so, over $A \cap A^*$, which is a subring invariant with respect to $*$. If $0 \neq s \in A \cap A^*$, where $s = s^*$, then $(s^{-1})^m \in A \cap A^*$ for some $m \geq 1$. Thus $s^{m-1}(s^{-1})^m \in A \cap A^*$, which is to say, $s^{-1} \in A \cap A^*$. If $0 \neq x \in A \cap A^*$ then $xx^* \in A \cap A^*$ and so $(xx^*)^{-1} \in A \cap A^*$; therefore $x^{-1} = x^*(xx^*)^{-1} \in A \cap A^*$. In short, if D is S-radical over A then we may assume that A _is a subdivision ring of_ D _such that_ $A = A^*$. _In all that follows we thus suppose that_ $A = A^*$ _is a subdivision ring of_ D _and that_ D _is S-radical over_ A, $A \neq D$.

LEMMA 6.2.2. If $N \not\subset Z$ then D is of characteristic p, $p \neq 0$, and if $s \in S$ then $s^{p^n} \in A$ for some $n = n(s) \geq 1$.

_Proof._ Since $N \not\subset Z$, S generates D, hence $S \not\subset A$. Let $Z^+ = Z \cap S$. If $s \in S$ and $s \notin A$ then $Z^+(s) \supsetneqq A \cap Z^+(s)$. Because every element in $Z^+(s)$ is symmetric, $Z^+(s)$ is radical over $A \cap Z^+(s)$. By a result of Kaplansky [11], char $D = p \neq 0$ and, either $Z^+(s)$ is purely inseparable over $A \cap Z^+(s)$ or $Z^+(s)$ is algebraic over a finite field.

In this second possibility, $Z^+$ itself must be algebraic over a finite field, and $s^{p^k} = s$ for some $k > 0$. Consider $C_D(s)$; if $a^* = a \in C_D(s)$ A, then as is symmetric and not in A. Moreover, if as is purely inseparable over A, then, since $a \in A$, we would have s purely inseparable over A, which would contradict $s^{p^k} = s$, $s \notin A$. Hence $(as)^{p^t} = as$ for some $t > 0$. We can find an $m > 0$ such that both $(as)^{p^m} = as$ and $s^{p^m} = s$. Since $as = sa$, we get $a^{p^m} = a$ for $a^* = a \in C_D(s) \cap A$. If $t^* = t \in C_D(s)$, then $t^r \in A \cap C_D(s)$ so is algebraic over a finite field by the above. Hence t is algebraic over a finite field for any symmetric element in $C_D(s)$. By Theorem 3.1.1, $C_D(s)$ is commutative. Since $s$ is algebraic over $Z$, we have that D is finite-dimensional over $Z$. Because $Z^+$, hence $Z$, is algebraic over a finite field, we then have that $D$ is algebraic over a finite field. By a celebrated theorem of Jacobson, D is then commutative, contradicting $N \subset Z$. Thus $s$ is purely inseparable over $A \cap Z^+(s)$, and so $s^{p^{n(s)}} \in A$.

Note that in the very special case when $D$ is finite-dimensional over $Z$, and $A \supset Z$ then we can say a little more. For, the degree of inseparability of any element over $Z$ is then bounded, hence the degree of inseparability of any $s \in S$ over $A$, which contains $Z$, is also bounded. By the previous lemma we then have that $s^{p^n} \in A$ for every $s \in S$ and a fixed integer $n$.

We now want to develop, in the context of division rings, the symmetric analog of the hypercenter [6]. We make the

DEFINITION. $W = \{x \in D \mid xs^n = s^n x,\ n = n(x, s) > 1, \text{ all } s \in S\}$.

Our aim is to show that quite generally (i.e., when $\dim_Z D \geq 4$) $W$ coincides with $Z$. Our first step in this direction is the finite-dimensional case.

LEMMA 6.2.3. If $D$ is finite-dimensional over its center, then $ws^2 = s^2 w$ for all $w \in W$, $s \in S$.

Proof. If $N \subset Z$, then since $s^2 \in N$, we certainly have $ws^2 = s^2 w$. Similarly, if $W \subset Z$ we have the result. So we suppose $N \not\subset Z$, $W \not\subset Z$.

Suppose then that $w \in W$, $w \notin Z$. If $A = C_D(w)$ then $A \neq D$, $A \supset Z$ and $D$ is S-radical over $A$. Since $N \not\subset Z$, by Lemma 6.2.2, $\operatorname{char} D = p \neq 0$ and $s^{p^{m(s)}} \in A$ for all $s \in S$. By the comment preceding this lemma, $s^{p^n} \in A$ for all $s \in S$ for some fixed $n$.

If $s^{p^n} \notin Z$ then $C_D(s^{p^n}) \neq D$ and both $s$ and $w$ are in $C_D(s^{p^n})$. By induction on $\dim_Z D$ we then get that $ws^2 = s^2w$. Hence, if $s^2w \neq ws^2$ for $s \in S$ then $s^{p^n} \in Z$.

Since $\dim_Z D < \infty$, $Z$ cannot be finite, hence $Z^+$ is infinite. Suppose $s^2w \neq ws^2$; then $s^{p^n} \in Z$. If $t = t^*$ and $t^2w = wt^2$ then $(s + \lambda t)^2 w \neq w(s + \lambda t)^2$ for an infinite number of $\lambda \in Z^+$, otherwise $s^2w = ws^2$ results. Hence $(s + \lambda t)^{p^n} \in Z$ for an infinite number of $\lambda$ in $Z^+$. By a vander Monde determinant argument we get $t^{p^n} \in Z$. Hence $D$ is S-radical over $Z$. By Theorem 3.2.1 we have that $N \subset Z$, and so $s^2 \in N \subset Z$ for all $s \in S$, whence $ws^2 = s^2w$.

We now can obtain the analog of the hypercenter result, namely,

THEOREM 6.2.1. If $N \not\subset Z$ then $W \subset Z$.

Proof. Let $w \in W$ and $s \in S$ and let $D_1$ be the subdivision ring of $D$ generated by $w, w^*$, and $s$. Since $w \in W$, $ws^k = s^kw$ for some $k \geq 1$, and so $w^*s^k = s^kw^*$, hence $s^k \in Z(D_1)$.

If $t^* = t \in C_{D_1}(s)$ then $t^mw = wt^m$ for some $m \geq 1$; therefore $t^m$ commutes with $s$, $w$ and $w^*$, so is in $Z(D_1)$. By Theorem 3.2.1, $C_{D_1}(s)$ is finite-dimensional over its center. However, since $s^k \in Z(D_1)$, in $D_1$, $s$ is algebraic over $Z(D_1)$. These give us that $D_1$ is finite-dimensional over its center. By Lemma 6.2.3, we have that $ws^2 = s^2w$. In other words, $ws^2 = s^2w$ for all $s \in S$. However, if $N \not\subset Z$, $\{s^2 \mid s \in S\}$ generates $D$. From this we have that $W \subset Z$.

COROLLARY. If $D$ is S-radical over $A$, $A \neq D$, and $N \not\subset Z$, then $Z(A) \subset Z$ and $C_D(A) \subset Z$.

Generally, in speaking about rings with involution, 2-torsion or characteristic 2 offers difficulties. In our present situation, strangely enough, char $D = 2$ is the easy case. We now dispose of it.

LEMMA 6.2.4. If char $D = 2$ and $D$ is S-radical over $A$, $A \neq D$, then $N \subset Z$.

<u>Proof.</u> Suppose that $N \not\subset Z$. By Lemma 6.2.2, if $s \in S$ then $s^{2^n} \in A$ for some $n \geq 0$. Since $S$ generates $D$, we can thus find an $s \in S$, $s \notin A$ with $s^2 \in A$. Hence $sAs = sAs^{-1}$ is a subdivision ring of $D$, invariant with respect to $*$. Similarly, $(1+s)A(1+s)$ and $s(1+s)As(1+s)$ are subdivision rings invariant with respect to $*$. Let $B = A \cap sAs \cap (1+s)A(1+s) \cap s(1+s)As(1+s)$; from $s^2 \in A$, $(1+s)^2 = 1+s^2 \in A$ we have that $B$ is normalized by both $s$ and $1+s$. As is easy then, $B$ must be <u>centralized</u> by $s$, that is, $B \subset C_D(s)$.

Let $B_1 = sAs \cap (1+s)A(1+s) \cap s(1+s)As(1+s)$; then $s^2 \in B_1$ (since $s^2 \in A$) and $B_1$ is S-radical over $B_1 \cap A = B \subset C_D(s)$. Hence $s^2$ commutes with a power of every symmetric element in $B_1$; by Theorem 6.2.1 we have that $s^4 \in Z(B_1)$. Using the form of $B_1$, we see that $s^4$ is in the center of $A \cap (1+s)A(1+s) \cap s(1+s)As(1+s)$. Let $B_2 = (1+s)A(1+s) \cap s(1+s)As(1+s)$; then $B_2$ is S-radical over $B_2 \cap A \subset C_D(s^4)$, hence $s^8 \in Z(B_2)$ by Theorem 6.2.1. This gives us that $s^8$ is in the center of $A \cap sAs$. Hence, since $sAs$ is S-radical over $A \cap sAs$, we have that $s^{16} \in Z(sAs)$ by Theorem 6.2.1, in other words, $s^{16} \in Z(A) \subset Z$ (by the corollary above).

If $s \in S$, $s \notin A$ then $s^{2^r} \in A$ for some minimal $r$, and so $(s^{2^{r-1}})^2 \in A$; by the above we get that $s^{2^t} \in Z$ for some $t \geq 0$. In short, every $s \in S$, $s \notin A$, is radical over $Z$.

Let $s \in S$, $s \notin A$; consider $C_D(s)$. $C_D(s)$ is S-radical over $A \cap C_D(s) \neq C_D(s)$, and every $t^* = t \in C_D(s)$, not in $A \cap C_D(s)$, has a power in $Z$. If $a = a^* \in A \cap C_D(s)$ then $as$ is symmetric and not in $A \cap C_D(s)$, whence $(as)^{2^k} \in Z$. These give that $a^{2^k} \in Z$, and so $C_D(s)$ is S-radical over $Z$, hence must be at most 4-dimensional over $Z(C_D(s))$. But $s$ is algebraic over $Z$. The net result is that $D$ must be finite-dimensional over $Z$.

If $Z^+ \not\subset A$, if $a = a^* \in A$ and $\lambda = \lambda^* \notin A$ then $\lambda a \notin A$, so, by the above is radical over $Z$. This would give that $a^{2^n} \in Z$ for $a \in A \cap S$. Hence we would have that $D$ is S-radical over $Z$. By Theorem 3.2.1 we would have $N \subset Z$ and we would be done.

Suppose then that $Z^+ \subset A$. Since $D$ is finite-dimensional over $Z$, $Z^+$ is infinite and every element in $D$ is purely inseparable over $A$ of bounded index of inseparability $2^k$.

Let $s \notin A$, $s \in S$ and $a = a^* \in A$. Then $a + \lambda s \notin A$ for all $\lambda \in Z^+$. Hence $(a+\lambda s)^{2^k} \in Z$. A vander Monde argument shows that $a^{2^k} \in Z$. But then, $D$ is S-radical over $Z$, whence $N \subset Z$ follows

With the case char $D = 2$ settled <u>we assume henceforth that</u> char $D \neq 2$.

LEMMA 6.2.5. If $D$ is S-radical over $A \neq D$, and $S \not\subset Z$ then $A$ must be infinite dimensional over $Z(A)$.

Proof. Suppose that $\dim_{Z(A)} A < \infty$. We already know that $Z(A) \subset Z$. Pick $A$ such that $D$ is S-radical over $A$ and such that $\dim_{Z(A)} A$ is minimal. If $u \in D$ is unitary then $D$ is S-radical over $uAu^{-1}$, hence over $A \cap uAu^{-1}$. Now, if $A \neq uAu^{-1}$, then since $Z(A) \subset Z$, $Z(A) \subset A \cap uAu^{-1}$, and so the dimension of $A \cap uAu^{-1}$ is smaller than $\dim_{Z(A)} A$, a contradiction. Hence $A = uAu^{-1}$ for all unitary elements in $D$.

By Theorem 6.1.1 we have that $A \subset Z$ if $\dim_Z D > 16$. By Theorem 3.2.1 we would conclude that $S \subset Z$. Hence we may suppose that $\dim_Z D \leq 16$. Also, as we remarked after the proof of Theorem 6.1.1, we may assume that * is of the first kind (otherwise we obtain $A \subset Z$ if $\dim_Z D > 4$). The proof of Theorem 6.1.1 showed that $[A, K] \subset A$. If $S \not\subset Z$, since $K$ generates $D$, if $A \not\subset Z$ we have $ak - ka \neq 0$ for some $a \in A$, $k \in K$. If $\lambda \neq 0 \in Z$ then $a(\lambda k) - (\lambda k)a = \lambda(ak - ka)$ is in $A$, giving us $\lambda \in A$. In other words, $Z \subset Z(A) \subset Z$, so $Z = Z(A)$. But then $D = A \otimes_Z C_D(A)$; by the corollary to Theorem 6.2.1 we know that $C_D(A) \subset Z$. The upshot is the contradiction $D = A$. With this the lemma is established.

We are now able to prove

THEOREM 6.2.2. If $D$ is S-radical over $A$, $A \neq D$, then $N \subset Z$ and $\dim_Z D \leq 4$.

Proof. By Lemma 6.2.4, we may assume that char $D \neq 2$. By Lemma 6.2.5 we may assume that $\dim_{Z(A)} A = \infty$; in particular, $A$ must have skew elements.

If $M = \{v \in A \mid vk^n = k^n v$, some $n \geq 1$, all $k \in A\}$ then $M$ is invariant re the unitaries in A. Because $\dim_{Z(A)} A = \infty$, by Theorem 6.1.1, $M = A$ or $M \subset Z(A) \subset Z$. If $M = A$, by Lemma 6.2.1, $\dim_{Z(A)} A \leq 4$. So $M \subset Z$. Hence there exists a skew element $k$ such that no power of $k$ is in $Z$. Since $uku^{-1} \notin A$ for some unitary element we may assume that $k \notin A$. Since $k^2 \in S$, $k^{2q} \in A$, $k^{2q} \notin Z$, for some $q$.

By the argument above, there exists an $a^* = -a$ in $A$ such that $k^{2q}$ commutes with no positive power of $a$ (for $M \subset Z$). Let $s = a^2$. The elements $u = (1-k)(1+k)^{-1}$, $v = (1+k)(1-k)^{-1}$, $w = (1-k-a)(1+k+a)^{-1}$ and $x = (1-k+a)(1+k-a)^{-1}$ are unitary, so there exists an integer $m \geq 1$ such that all of $us^m u^{-1}$, $vs^m v^{-1}$, $ws^m w^{-1}$ and $xs^m x^{-1}$ are in A. These give (as in the proof of Theorem 6.1.1) that $b - kbk \in A$, $b - (k+a)b(k+a) \in A$ and $b - (k-a)b(k-a) \in A$ where $b = (ks^m - s^m k)^{-1}$. These relations lead us easily to $aba \in A$, and so $b \in A$. Hence $b^{-1} = ks^m - s^m k \in A$. By the same token — and we can arrange for the same $m$ — $k^{2q+1}s^m - s^m k^{2q+1} \in A$. But $A \ni k^{2q+1}s^m - s^m k^{2q+1} = (k^{2q}s^m - s^m k^{2q})k + k^{2q}(ks^m - s^m k)$. Now $k^{2q} \in A$, $ks^m - s^m k \in A$ and $k^{2q}s^m - s^m k^{2q} = k^{2q}a^{2m} - a^{2m}k^{2q} \neq 0 \in A$; Thus we get the contradiction $k \in A$. With this the proof is complete.

The theorem proved above allows us to prove a type of commutativity theorem. In division rings D (and, more generally, in rings with no nil ideals [1], [7]) if $x^m y^n = y^n x^m$, $m = m(x, y) \geq 1$, $n = n(x,y) \geq 1$ for every x,y in D, then D is a field. We show that if we restrict the condition to symmetric elements, then D must be rather special. This is

THEOREM 6.2.3. Let D be a division ring with involution in which, for any $a, b \in S$, there exist integers $m = m(a, b) \geq 1$, such that $a^m b^n = b^n a^m$. Then $N \subset Z$ and $\dim_Z D \leq 4$.

Proof. If $N \not\subset Z$, by Lemma 6.2.1 there is an $a \in S$ such that $A = \{x \in D \mid xa^r = a^r x \text{ some } r \geq 1\}$ is not all of D. However, by hypothesis, D is S-radical over A. Thereorem 6.2.2 then forces the contradiction $N \subset Z$. This proves the theorem.

Analogous theorems exist for the skew elements. We state these theorems here but do not go into their proofs. These can be found in [2]. While the proofs require some techniques which are different from the ones we used for S, by and large, the pattern is very similar.

Say that D is K-radical over $A \neq D$ if $k^{n(k)} \in A$ for all $k \in K$. Then:

If D is K-radical over $A \neq D$ then $k^2 \in Z$ for all $k \in K$, and $\dim_Z D \leq 4$.

From this one shows: if, given $a, b \in K$, $a^m b^n = b^n a^m$ for some $m = m(a, b) \geq 1$, $n = n(a, b) \geq 1$ then $a^2 \in Z$ for all $a \in K$, and $\dim_Z D \leq 4$.

## 3. K-invariant Subrings

In this section we continue the study of dichotomy of the Brauer-Cartan-Hua kind that has come up in so much of the work on rings with involution. The results we prove are closely related to those of Section 1 of this chapter; in fact, in the proof of the complete Theorem 6.1.1 in [5], use is made of Theorem 6.3.1.

While characteristic 2 usually offers difficulties in studying rings with involution, and dimensions 4 and 16 often play the role of counter-examples, in Theorem 6.3.1 we have a new villain — the $3 \times 3$ matrices in characteristic 3 provide a counter-example to the theorem in question.

We shall be concerned with subrings, A, of a simple ring R which are invariant with respect to commutation with skew elements, that is, $[A, K] \subset A$. Our general objective is to show that $A \subset Z$ or $A = R$. When $\dim_Z R = 4$ or $\dim_Z R = 16$ this may be false. However here we have a new twist. In $R = F_3$, where F is a field of characteristic 3, the subring

$$A = \left\{ \begin{pmatrix} \alpha & \beta & \gamma \\ \beta & \alpha+\beta-\gamma & -\beta-\gamma \\ \gamma & -\beta-\gamma & \alpha+\gamma-\beta \end{pmatrix} \,\middle|\, \alpha, \beta, \gamma \in F \right\}$$

is a commutative subring, clearly not in $Z(R) = F$, which consists of symmetric elements — relative to transpose — such that $[A, K] \subset A$.

The results of this section come from [8].

LEMMA 6.3.1. Let R be a simple ring with involution of the second kind, char $R \neq 2$. Suppose that A is a commutative set of elements of R such that $[A, K] \subset A$. Then $A \subset Z$.

Proof. The subring generated by A satisfies the same hypotheses as does A, hence we may assume, without loss of generality, that A is a subring of R, and is, in fact, a subalgebra over the centroid C of R.

Since $*$ is of the second kind, there is a $\lambda \neq 0$ in C such that $\lambda^* = -\lambda$; hence $S = \lambda K$. Now $[A, S] = [A, \lambda K] = \lambda[A, K] \subset \lambda A \subset A$, thus, since $R = S + K$, $[A, R] \subset A$. Since $A$ is a Lie ideal of $R$, and is a commutative subring, by Theorem 2.1.3 we have that $A \subset Z$.

In all that follows we may therefore assume that * is of the first kind.

The key to our first theorem, Theorem 6.3.1, is the result

LEMMA 6.3.2. Let $R$ be simple with involution, char $R \neq 2$. Suppose that $A$ is a commutative set of symmetric elements such that $[A, K] \subset A$. Then:

(1) if char $R \neq 3$ and $\dim_Z R > 4$, $A \subset Z$,

(2) if char $R = 3$ and $\dim_Z R > 9$, $A \subset Z$.

Proof. The subalgebra generated by $A$ over the centroid $C$ of $R$ satisfies the same hypotheses as $A$, since * is of the first kind. Furthermore, the conditions carry over to $R \otimes_C F$ and $A \otimes_C F$ where $F$ is the algebraic closure of $C$. Hence we may assume that $C$ is algebraically closed.

If $a \in A$, let $d(x) = ax - xa$ for $x \in R$. Our hypothesis on $A$ implies that $d^2(k) = 0$ for every $k \in K$. Also, since $a \in A$ is symmetric, if $s \in S$ then $d(s) \in K$ hence $d^3(s) = 0$. Since $R = S + K$, we get $d^3(x) = 0$ for all $x \in R$. This translates into $a^3x - 3a^2xa + 3axa^2 - xa^3 = 0$, hence $a^3R \subset Ra$ and $Ra^3 \subset aR$. Because $R$ is simple this gives that $a^3 = 0$ or $a$ is invertible in $R$. We now divide the argument according as char $R \neq 3$ or char $R = 3$.

If char $R \neq 3$ and $a \in R$ is not invertible, then $a^3 = 0$ hence from $d^3(x) = 0$ we get $3a^2xa = 3axa^2$ and so $a^2xa = axa^2$. This yields $0 = a^3xa = a^2xa^2$; since $R$ is simple, we have $a^2 = 0$. Thus any nilpotent element in $A$ has square 0. If $a, b \in A$ are nilpotent then so is $a+b$, since they commute, whence $0 = (a+b)^2 = a^2 + 2ab + b^2 = 2ab$, resulting in $ab = 0$.

If $d^2(x) = 0$ for all $x \in R$ then $a$ commutes with all its commutators; by Lemma 1.1.9 we have $a \in Z$. So we may suppose that $d^2(x) \neq 0$ for some $x \in R$. If $k \in K$ then $d^3(xk) = 0$; expanding by Leibniz' rule and using $d^3(x) = d^2(k) = 0$ we get $3d^2(x)d(k) = 0$ and so $d^2(x)d(k) = 0$. Since $d(k) \in A$ is thus a zero-divisor it is not invertible; by the above, $d(k)^2 = 0$. Hence $(ak - ka)^2 = 0$. Now $a(ak - ka) = (ak - ka)a$ yields, if $a^2 = 0$, that $aKa = 0$. Thus $(ak - ka)^2 = 0$ gives us $ak^2a = 0$ for every $k \in K$. But by Baxter's theorem (Theorem 2.1.11), the linear span of the $k^2$, $k \in K$, in S; therefore $aSa = 0$. From $aKa = aSa = 0$ we get $aRa = 0$, and so $aRa = 0$. In other words, $A$ has no nilpotent elements. But for any $a \in A$, $ak - ka$ is nilpotent if $k \in K$; thus $ak = ka$ for all $k \in K$. Since $\dim_Z R > 4$, $K$ generates $R$. The upshot is that $a \in Z$ and so $A \subset Z$ results. This finishes char $R \neq 3$.

Suppose that char $R = 3$. If $a \in A$, from $d^3(x) = 0$ we get $a^3x = xa^3$, and so $a^3 \in Z$. If $0 \neq b \in A$ and $b^2 = 0$ we saw before that from $b(bk - kb) = (bk - kb)b$ follows $bKb = 0$. Hence, since $x - x^* \in K$ for all $x \in R$, $bxb = bx^*b$. If $c \in A$ then $c^* = c$, $bc = cb$, and $b(cx)b = b(cx)^*b = bx^*cb = bx^*bc = bxbc$. Thus, by Lemma 1.3.2, $bc = \lambda(c)b$ where $\lambda(c) \in C$. If $c$ is nilpotent this gives $\lambda(c) = 0$, and so $bc = 0$. Not every element in $A$ can then be nilpotent, otherwise, for any $k \in K$, $0 = b((bk - kb)k - k(bk - kb)) = bk^2b$ follows; as before we get $bSb = 0$ and so $bRb = 0$, whence the contradiction $b = 0$. So, $A$ has non-nilpotent elements, hence an invertible element, whence $R$ has a unit element and $C = Z$.

If $0 \neq b \in A$ and $b^2 = 0$, we saw that $bxb = bx^*b$ for all $x \in R$; using $x = ubv$ we get $bubvb = bvbub$ for all $u, v \in R$. By Theorem 1.3.1, $R$ has a minimal right ideal and the commuting ring of $R$ must be a field. Since $1 \in R$ and $R$ is simple with minimal right ideal, $R$ is artinian. Because the commuting ring of $R$ is a field, $R = F_n$ for some field $F$.

We claim that $n \leq 3$. For, if $b^2 = 0$, $b \neq 0 \in A$, we saw $bk^2b \neq 0$ for some $k \in K$, and so $bc \neq 0$ where $c = (bk - kb)k - k(bk - kb)$. Thus $c = bk^2 + kbk + k^2b$ is not nilpotent; being in $A$, $c$ is thus invertible. Finally, from $bubvb = bvbub$ for all $u, v \in R$ we have, by Theorem 1.3.1, that $bub = \lambda(u)b$, $\lambda(u) \in F$; hence rank $b = 1$. But then, rank $c \leq 3$; since $c$ is invertible, rank $c = n$. In consequence, $n \leq 3$, contrary to $\dim_Z R > 9$.

In short, $A$ <u>cannot</u> have nilpotent elements; hence every element in $A$ is invertible, whence $1 \in R$ and $Z = C$ is algebraically closed. If $a \neq 0 \in A$ then $a^3 = \mu \neq 0$ where $\mu \in Z$; thus $Z \subset A$ follows. Also, $\mu = \nu^3$ for some $\nu \in Z$ since $Z$ is algebraically closed. Thus $(a - \nu)^3 = 0$. But $a - \nu \in A$ is nilpotent, whence $a - \nu = 0$. Therefore $a = \nu$ and we have shown that $A \subset Z$.

We are ready to move on to the main result of this section,

THEOREM 6.3.1. Let $R$ be a simple ring with involution, char $R \neq 2$. Suppose that $A$ is a subring of $R$ such that $[A, K] \subset A$. Then:

1. if $A$ is not commutative and $\dim_Z R > 16$, $A = R$.
2. if $A$ is commutative, $\dim_Z R > 4$ and char $R \neq 3$, $A \subset Z$.
3. if $A$ is commutative, char $R = 3$ and $\dim_Z R > 9$, $A \subset Z$.

Proof. We first dispose of the case $A = A^*$. In this case $A$ must have non-symmetric elements, otherwise the result follows directly from Lemma 6.3.2. Hence $A$ must have skew elements; thus $A^- = A \cap K$ is not 0. However, $A^-$ is a Lie ideal of $K$ since $[A, K] \subset A$; if $A^- \subset Z$ and $\lambda \neq 0 \in A^-$ then $\lambda s \in A^-$ for every $s \in A \cap S$, whence $s \in Z$ follows. This gives $A \subset Z$, since $A = A \cap K \oplus A \cap S$.

We may suppose that $A^- \not\subset Z$. If $A$ is commutative and $\dim_Z R > 4$, if $a \neq 0 \in A^-$ then $a^2k - ka^2 \in A$ and $a^2s - sa^2 = a(as + sa) - (as + sa)a \in A$ since $as + sa \in K$. Therefore $a^2x - xa^2 \in A$ for all $x \in R$, hence commutes with $a^2$. By Lemma 1.1.9, $a^2 \in Z$. But $0 = a(ak - ka) + (ak - ka)a = 2a(ak - ka)$ then yields, if $a^2 \neq 0$, that $ak = ka$; since $K$ generates $R$ we would have $a \in Z$. On the other hand, if $a^2 = 0$ then $a(ak - ka) = 0$ yields $aKa = 0$. If $s \in S$ then $sas \in K$ and so $asasa = 0$. Therefore $(ax)^3 = 0$ for all $x \in R$; by Lemma 2.1.1, $a = 0$. These things contradict $A^- \not\subset Z$, and so we have the result if $A$ is commutative.

If $A$ is not commutative and $\dim_Z R > 16$, since $A^-$ is a non-central Lie ideal of $K$, $A^- \supset [K, K]$ by Theorem 2.12 of TRT. But by

Theorem 2.13 of TRT, $[K, K]$ generates $R$. This gives $A = R$. We have finished the case $A = A^*$.

We go to the general case. If $B = A \cap A^*$ then $[B, K] \subset B$ and $B^* = B$. If $A \neq R$ then $B \neq R$; by the discussion above, $B \subset Z$.

If $a, b \in A$ then $ab^* - b^*a = a(b^* - b) - (b^* - b)a + ab - ba$ is in $A$; however, $ab^* - b^*a = (a - a^*)b^* - b^*(a - a^*) + a^*b^* - b^*a^*$ is also in $A^*$. Thus $ab^* - b^*a \in B \subset Z$ for all $a, b \in A$. Therefore $a^2b^* - b^*a^2 = 2a(ab^* - b^*a)$ is also in $Z$; if $ab^* - b^*a \neq 0$, since it is in $Z$ and $a(ab^* - b^*a)$ is in $Z$ we would have $a \in Z$. This would give $ab^* - b^*a = 0$, a contradiction. Hence we have $ab^* = b^*a$ for all $a, b \in A$.

Let $C = A + A^* + AA^*$; from the above, $C$ is a subring of $R$, $C^* = C$ and $[C, K] \subset C$. If $A$ is commutative then $C$ is commutative; so under the appropriate condition on $\dim_Z R$ and char $R$ we get $C \subset Z$, whence $A \subset Z$. Thus we may assume that $A$ is not commutative and $\dim_Z R > 16$. By our previous discussion, $C = R$.

Let $a, b \in A$ such that $ab - ba \neq 0$. Because $k = a^*b - b^*a \in K$, $ak - ka \in A$; but $ak - ka = a^*(ab - ba)$, so $a^*(ab - ba) \in A$. If $c \in A$ we must have $c^*$ commuting with $a^*(ab - ba)$, giving us $(a^*c^* - c^*a^*)(ab - ba) = 0$.

Because $R = A + A^* + AA^*$, given $x \in R$, $x = a_1 + a_2^* + \sum u_i v_i^*$ where $a_1, a_2$ and all the $u_i$ and $v_i$ are in $A$. Thus $(a^*x - xa^*)(ab - ba) = (a^*a_2^* - a_2^*a^*)(ab - ba) + \sum u_i(a^*v_i^* - v_i^*a^*)(ab - ba) = 0$ from the above. Since $ab - ba \neq 0$ and annihilates all commutators of $a^*$, by Lemma 1.1.7 $a^* \in Z$, whence $a \in Z$. This contradicts $ab - ba \neq 0$. The proof is now complete.

We now propose to investigate subrings in simple rings invariant with respect to particular combinations of symmetric or skew elements. The proofs of these theorems are considerably easier than that of the theorem we just proved.

THEOREM 6.3.2. Let $R$ be a simple ring with involution, char $R \neq 2$, and $\dim_Z R > 4$. If $A$ is an additive subgroup of $R$ such that $[A, S] \subset A$ then either $A \subset Z$ or $A \supset [R, R]$. Furthermore, if $A$ is a subring of $R$ then $A \subset Z$ or $A = R$.

Proof. By Lemma 4.1.1 we know that $[S, S] \supset [K, K]$. Because $R = S + K$, $[R, R] = [S, S] + [S, K] + [K, K]$, thus $[R, R] \subset [S, S] + [K, S]$. But $[R, S] = [S + K, S] = [S, S] + [K, S]$, therefore $[R, R] \subset [R, S]$. Since trivially, $[R, S] \subset [R, R]$, we have $[R, R] = [R, S]$.

We now prove the theorem. Since $[A, S] \subset A$, by the Jacobi identity we get that $[A, [S, S]] \subset A$. Thus, since $[R, R] = [R, S] = [S, S] + [S, K] \subset S + [S, S]$, $[A, [R, R]] \subset [A, S] + [A, [S, S]] \quad A$. By Theorem 2.1.4 we conclude that $A \subset Z$ or $A \supset [R, R]$. If $A$ is a subring, because $[R, R]$ generates $R$ we get, if $A \supset [R, R]$, that $A = R$.

We now take a look at subsystems invariant to the circle product, $a \circ b = ab + ba$, relative to $K$ and $S$. The first of these results is

THEOREM 6.3.3. Let $R$ be a simple ring with involution, char $R \neq 2$, and $\dim_Z R > 4$. If $A$ is an additive subgroup of $R$ such that $A \circ K \subset A$ then $A = 0$ or $A = R$.

Proof. Suppose that $A \neq 0$. If $a \in A$ and $k \in K$ then $(ak + ka)k + k(ak + ka) = ak^2 + 2kak + k^2a$ is in $A$. We linearize this on $k$ to obtain

(1) $a(k_1k_2 + k_2k_1) + (k_1k_2 + k_2k_1)a + 2k_1ak_2 + 2k_2ak_1 \in A$ for $a \in A$, $k_1, k_2 \in K$.

But $k_1k_2 - k_2k_1 \in K$, hence

(2) $a(k_1k_2 - k_2k_1) + (k_1k_2 - k_2k_1)a \in A$.

Using that $2K = K$, adding (1) and (2) we get

(3) $ak_1k_2 + k_1k_2a + k_1ak_2 + k_2ak_1 \in A$.

However, $(ak_1 + k_1a)k_2 + k_2(ak_1 + k_1a) \in A$; subtracting this from (3) we have $(k_1k_2 - k_2k_1)a \in A$, that is, $[K, K]A \subset A$. Since $\dim_Z R > 4$, $[K, K]$ generates $R$; thus from $[K, K]A \subset A$ we get $RA \subset A$. Now $(RA) \circ K \subset A$; since $RA \subset A$ this leads to $RAK \subset A$. Since $K$ generates $R$, and $R^2AK = RAK \subset RA$ we get $RAR = A$. Since $A \neq 0$ and $R$ is simple, $RAR = A$. Consequently, $A = R$ follows.

We close this section with the last result of this type.

THEOREM 6.3.4. Let $R$ be a simple ring with involution such that char $R \neq 2$ and $\dim_Z R > 4$. If $A$ is a subring of $R$ such that $A \circ S \subset A$ then $A = 0$ or $A = R$.

<u>Proof.</u> If $a, b \in A$ and $s \in S$ then $abs - sab = a(bs + sb) - (as + sa)b \in A^2$. Hence, $[A^2, S] \subset A^2$. By Theorem 6.3.3, either $A^2 \subset Z$ or $A^2 \supset [R, R]$.

If $A^2 \supset [R, R]$, since $[R, R]$ generates $R$, we have that $A = R$.

Suppose, then, that $A^2 \subset Z$. If $A^2 \neq 0$, let $\lambda \neq 0 \in A^2 \subset Z$. If $s \in S$ then $\lambda s + s\lambda = 2\lambda s$ is in $A$, so $\lambda S \subset A$. Hence $(\lambda s)^2 \in A^2 \subset Z$ gives that $s^2 \in Z$. But $\{s^2 \mid s \in S\}$ generates $R$, since $\dim_Z R > 4$. This contradicts $s^2 \in Z$. We must have $A^2 = 0$.

If $a, b \in A$ and $s \in S$ then $ab = 0$ and $a(bs + sb) = 0$, therefore $asb = 0$. Let $W = \{x \in R \mid axb = 0, \text{ all } a, b \in A\}$; the remark above shows that $W \supset S$, hence $W \neq 0$.

If $x \in W$ and $s \in S$ then, for $a, b \in A$, $ax(bs + sb) = 0$ and $(as + sa)xb = 0$. These give $axsb = 0$ and $asxb = 0$; in other words, $WS \subset W$ and $SW \subset W$. Since $S$ generates $R$ we have $WR \subset W$ and $RW \subset W$, whence $RWR \subset RW \subset W$. Because $W \neq 0$ and $R$ is simple, $RWR = R$, hence $W = R$. The very definition of $W$ then tells us that $aRb = 0$ for all $a, b \in A$. By the simplicity of $R$ we conclude that $A = 0$.

## 4. Another Dichotomy Theorem

We shall consider rings with involution and certain subrings thereof which are invariant with respect to a particular kind of combination of elements. Our aim — as is usually the case for theorems of this sort — will be to show that the subrings are "small", namely must be central, or are "large" in the sense that they contain non-zero ideals of the ring.

In everything we do in this section, $R$ will be a semi-prime ring with involution $*$, and $A$ will be a subring of $R$ such that $xAx^* \subset A$ for all $x \in R$.

We want to show that $A \subset Z = Z(R)$ or that $A$ must contain a non-zero ideal of $R$. When $A \subset Z$ and $A$ contains no non-zero ideal of $R$ we can say quite a bit more, namely that $R$ must satisfy the standard identity in 4 variables when it is $*$-prime.

We shall show the result by a series of reductions, reducing down to the case in which $A$ is a commutative subring consisting of symmetric elements. This accounts for the first lemma that we prove.

Before starting the proof proper we make a remark about prime rings which we could have made earlier. Let $R$ be a prime ring and suppose that $U \neq 0$ is an ideal of $R$; suppose that $U$ satisfies a multilinear, homogeneous polynomial identity $p(x_1, \ldots, x_n)$. Then $R$ satisfies $p(x_1, \ldots, x_n)$. To see this, note that since $U$ is prime, by Formanek's theorem (in fact, by Theorem 1.4.2), $Z(U) \neq 0$. Moreover, by Lemma 1.1.5, $Z(U) \subset Z(R)$. If $\lambda \neq 0 \in Z(U)$ and $r_1, \ldots, r_n \in R$ then, since $\lambda r_1, \ldots, \lambda r_n$ are in $U$, $p(\lambda r_1, \ldots, \lambda r_n) = 0$. But $0 = p(\lambda r_1, \ldots, r_n) = \lambda^n p(r_1, \ldots, r_n)$ since $\lambda \in Z(R)$; because elements in $Z(R)$ are not zero divisors in $R$, we get $p(r_1, \ldots, r_n) = 0$, thereby proving our assertion.

LEMMA 6.4.1. Suppose that $R$ is semi-prime and that $A$ is a subring of $R$ consisting of symmetric elements such that $xAx^* \subset A$ for all $x \in R$. Then $A \subset Z$.

Proof. Since $R$ is semi-prime it is a subdirect product of prime rings $R_\alpha$; if we could show that the image of $A$ in each $R_\alpha$ lies in $Z(R_\alpha)$ then we would have that $A \subset Z(R)$.

If $P$ is a prime ideal of $R$ and $P^* \neq P$, consider $\overline{R} = R/P^*$; $\overline{W} = \dfrac{P + P^*}{P^*}$ is a non-zero ideal of $\overline{R}$. What can we say about $\overline{A} = \dfrac{A + P^*}{P^*}$? Linearizing $xax^* \in A$ for $a \in A$, $x \in R$ gives us that $xay + y^*ax^* \in A$ for all $x, y \in R$. Hence in $\overline{R}$, if $x, y \in P$ we get $\overline{W}\,\overline{A}\,\overline{W} \subset \overline{A}$. Because $A$ consists of symmetric elements $A$ must be commutative, hence $\overline{A}$ is commutative. Thus $\overline{WAW}$ is a non-zero commutative ideal of $\overline{R}$ which is prime; the upshot of this is that $\overline{R}$ is commutative. In this case $\overline{A}$ is certainly contained in $Z(\overline{R})$.

So, to finish the proof (because we need but consider $P^* = P$) we may assume that $R$ is prime. The linearized version of $xAx^* \subset A$ gives us that $xay + y^*ax^* \in A$ for all $x, y \in R$; thus, if $U = RaR$ then, since $a^* = a$, $u + u^* \in A$ for all $u \in U$. Thus any two traces in U, being in A, must commute. By Amitsur's theorem (Theorem 5.1.2) U must satisfy a standard identity, hence by the remark we made above, R satisfies this standard identity. Therefore $Z(R) \neq 0$ and the localization, Q, of R at $Z(R)$ is a finite-dimensional simple algebra over a field F (F the field of quotients of $Z(R)$). If B is the localization of A at Z, since we saw that $u + u^* \in A$ for all $u \in RaR = U$ and since U localizes to Q, we get that all traces in Q are in the commutative ring B.

If char $R \neq 2$ this tells us that any two symmetric elements of Q, hence of R, must commute. By Theorem 2.1.6, all symmetrics are in $Z(Q)$, whence $A \subset Z$ since A consists of symmetric elements. Similarly, if $*$ is of the second kind on $Z(R)$, if $\lambda \neq \lambda^* \in Z(R)$ then $\lambda s + (\lambda s)^* = (\lambda + \lambda^*)s \in B$ for $s \in S$, whence, if char $R = 2$, we get any two symmetrics of R commute, so again by Theorem 2.1.6, we get $A \subset Z$.

Thus we may assume that char $R = 2$ and $*$ is of the first kind. From the linearized form of $xAx^* \subset A$ we easily get that $uBu^* \subset B$ for all $u \in Q$, and that this persists for $B \otimes_F \overline{F}$ in $Q \otimes_F \overline{F}$, where $\overline{F}$ is the algebraic closure of F. Thus we may assume that F is algebraically closed. Since all traces in Q commute we have that $Q = F_2$.

Thus $*$ on Q is either transpose or the symplectic involution. If $*$ is transpose then, since $B \supset \{x + x^* = x \in Q\}$ we have $\begin{pmatrix} 0 & 1 \\ 1 & 0 \end{pmatrix} \in B$; thus $\begin{pmatrix} 1 & 0 \\ 0 & 0 \end{pmatrix} = \begin{pmatrix} 0 & 1 \\ 1 & 0 \end{pmatrix}^2 \in B$ and so

$\begin{pmatrix} 1 & 1 \\ 0 & 1 \end{pmatrix}\begin{pmatrix} 1 & 0 \\ 0 & 1 \end{pmatrix}\begin{pmatrix} 1 & 0 \\ 1 & 1 \end{pmatrix} = \begin{pmatrix} 0 & 1 \\ 1 & 1 \end{pmatrix} \in B$. We thus get $\begin{pmatrix} 0 & 0 \\ 0 & 1 \end{pmatrix} \in B$.

$\begin{pmatrix} 0 & 0 \\ 0 & 1 \end{pmatrix}\begin{pmatrix} 0 & 1 \\ 1 & 0 \end{pmatrix} \neq \begin{pmatrix} 0 & 1 \\ 1 & 0 \end{pmatrix}\begin{pmatrix} 0 & 0 \\ 0 & 1 \end{pmatrix}$, contradicting that B is commutative. So the transpose case cannot arise. We may thus suppose that $\begin{pmatrix} \alpha & \beta \\ \gamma & \delta \end{pmatrix} = \begin{pmatrix} \delta & \beta \\ \gamma & \alpha \end{pmatrix}$ for $\begin{pmatrix} \alpha & \beta \\ \gamma & \delta \end{pmatrix} \in Q$.

Suppose that $\begin{pmatrix} \alpha & \beta \\ \gamma & \alpha \end{pmatrix} \in B$ where $\gamma \neq 0$; then

$\begin{pmatrix} 0 & 1 \\ 0 & 0 \end{pmatrix}\begin{pmatrix} \alpha & \beta \\ \gamma & \alpha \end{pmatrix}\begin{pmatrix} 0 & 1 \\ 0 & 0 \end{pmatrix} \in B$, that is, $\begin{pmatrix} 0 & \gamma \\ 0 & 0 \end{pmatrix} \in B$. Because $\gamma \neq 0$, we have $\begin{pmatrix} 0 & 1 \\ 0 & 0 \end{pmatrix} \in B$. Thus $\begin{pmatrix} 0 & 0 \\ 1 & 0 \end{pmatrix}\begin{pmatrix} 0 & 1 \\ 0 & 0 \end{pmatrix}\begin{pmatrix} 0 & 0 \\ 1 & 0 \end{pmatrix} = \begin{pmatrix} 0 & 0 \\ 1 & 0 \end{pmatrix} \in B$. However, $\begin{pmatrix} 0 & 1 \\ 0 & 0 \end{pmatrix}\begin{pmatrix} 0 & 0 \\ 1 & 0 \end{pmatrix} \neq \begin{pmatrix} 0 & 0 \\ 1 & 0 \end{pmatrix}\begin{pmatrix} 0 & 1 \\ 0 & 0 \end{pmatrix}$, in consequence of which $\gamma = 0$. Similarly, $\beta = 0$. Thus every element of B is a scalar, so $B \subset Z(Q)$ and $A \subset Z(R)$. The lemma is proved.

We now start the reduction process of the general result to the situation of Lemma 6.4.1. The first step is

LEMMA 6.4.2. If A does not contain a non-zero ideal of R then $ab^* = ab$ for all $a, b \in A$.

<u>Proof.</u> By linearizing $xax^* \in A$ we get

(1) $\qquad xay^* + yax^* \in A$ for $a \in A$, $x, y \in R$.

In (1) replace $x$ by $xb$ where $b \in A$; we get

(2) $$xbay^* + yab^*x^* \in A.$$

But, since $ba \in A$, by (1) we have

(3) $$xbay^* + ybax^* \in A.$$

Subtracting (3) from (2) yields $y(ab^* - ba)x^* \in A$ for all $x, y \in R$, hence $R(ab^* - ba)R \subset A$. Because $A$ contains no ideal of $R$, we conclude that $R(ab^* - ba)R = 0$ and so, $ab^* - ba = 0$.

A similar argument, replacing $y$ by $yb^*$ gives us that $b^*a = ab$.

Lemma 6.4.2 allows us to settle the result when $A$ is non-commutative.

LEMMA 6.4.3. If $A$ is not commutative then $A$ contains a non-zero ideal of $R$.

Proof. Suppose the result is false; by Lemma 6.4.2, $ab^* = ba$ for all $a, b \in A$. If $a, b, c \in A$ then $ab^*c^* = bac^* = bca$. However, $ab^*c^* = a(cb)^* = cba$. The net result is that $(bc - cb)A = 0$; if $W = \{w \in R \mid wA = 0\}$ then, since $[A, A] \subset W$ and $A$ is not commutative we must have that $W \neq 0$.

If $0 \neq w \in W$, $x \in R$, and $a, u \in A$ then since $xau^* + uax^* \in A$, $w(xau^* + uax) = 0$. However, $wu = 0$ since $u \in A$; thus $wxau^* = 0$, that is, $wRau^* = 0$, which is to say $WRAA^* = 0$, and so $WRAA^*R = 0$. Now $cb^* = bc$ and $bc^* = cb$, therefore $W \ni bc - cb = cb^* - bc^*$, and $cb^* - bc^* \in AA^*$. This gives $((bc - cb)R)^2 \subset WRAA^*R = 0$. Since $R$

is semi-prime we get $bc - cb = 0$, contradicting $bc - cb \neq 0$. With this the lemma is proved.

We now carry out the final step of the reduction.

LEMMA 6.4.4. If $A$ is commutative and $A$ does not contain a non-zero ideal of $R$ then every element in $A$ is symmetric.

Proof. By Lemma 6.4.2, $ab^* = ba$ and $b^*a = ab$ for $a, b \in A$. By Lemma 6.4.3, $ab = ba$; therefore $(b^* - b)a = 0$ and $b^*a = ab^*$. Therefore $A$ centralizes $A^*$ and $(b^* - b)A = 0$ for all $b \in A$. Applying $*$ to $(b^* - b)A = 0$ we get, since $b^* - b$ centralizes $A^*$, that $(b^* - b)A^* = 0$. Let $t = b^* - b$.

If $x \in R$ then $xtx^* = xb^*x^* - xbx^* \in A + A^*$ hence $txtx^* \in tA + tA^* = 0$. Similarly, $x^*txt = 0$. Linearize $0 = txtx^*$ on $x$; we get $txty^* + tytx^* = 0$ for $x, y \in R$. Multiply this last relation from the right by $txt$; since $x^*txt = 0$, we get $txty^*txt = 0$, that is, $txtRtxt = 0$. However, $R$ is semi-prime, hence $txt = 0$ for all $x \in R$. Thus $tRt = 0$. Again, by the semi-primeness of $R$, $t = 0$ results, which is to say, $b^* = b$. Hence every element in $A$ is symmetric.

We have all the pieces to prove our first theorem. This is from [9].

THEOREM 6.4.1. If $R$ is a semi-prime ring with involution $*$ and $A$ is a subring of $R$ such that $xAx^* \subset A$ for all $x \in R$, then either $A \subset Z$ or $A$ contains a non-zero ideal of $R$.

We conclude this section with a slight sharpening of the result for $*$-prime rings.

THEOREM 6.4.2. If $R$ is $*$-prime and $A \neq 0$ is a subring of $R$ such that $xAx^* \subset A$ for all $x \in R$, then, if $A$ contains no non-zero ideal of $R$, $A \subset Z$ and $R$ satisfies the standard identity of degree 4.

Proof. As the proof of Theorem 6.4.1 showed, since $A$ contains no non-zero ideal of $R$, $A \subset Z$ and $A$ consists of symmetric elements. If $a \neq 0 \in A$ then $xax^* = axx^*$ is in $A$, hence in $Z$, for all $x \in R$. Since $R$ is $*$-prime and $a$ is a non-zero symmetric element in $Z$, $a$ is a non-zero divisor. This shows that $xx^* \in Z$ for all $x \in R$; thus $xxx^* = xx^*x$, whence $x(xx^* - x^*x) = 0$. Since $xx^* - x^*x$ is a symmetric element in $Z$, from $x(xx^* - x^*x) = 0$ we conclude that $xx^* = x^*x$. From this point, either by quoting Amitsur's theorem (Theorem 5.1.1) or by a direct simple computation, we get that $R$ satisfies the identities of the $2 \times 2$ matrices over a field, hence the standard identity in 4 variables.

## 5. Relations Between $R$ and $\overline{S}$ or $\overline{K}$

If $R$ is a ring with involution and $\overline{S}$ and $\overline{K}$ are, respectively, the subrings of $R$ generated by $S$ and $K$, then we have seen earlier some relationship between the structure of $R$ and those of $\overline{S}$ and $\overline{K}$. For instance, if $R$ is simple, then except for very low dimensional cases (namely, $\dim_Z R \leq 4$) $\overline{S}$ and $\overline{K}$ are both equal to $R$, hence are simple. We shall use these earlier results in what is to follow; the net result will be a broad generalization of these earlier things.

In this section we shall examine the inter-relation between $\overline{S}, \overline{K}$ and $R$ more systematically and a little deeper. The results come mostly from the work of Lanski [12, 13] and Lee [14]. However, the proofs here are completely different from theirs and are considerably easier. Lee worked in a slightly more general context, that of a symmetric subring $U$ which he defined by:

1. $U$ is generated by some subset of $S$
2. all $x+x^*$ and $xx^*$ are in $U$
3. $yUy^* \subset U$ for all $y \in R$.

The arguments we shall give for $\overline{S}$ work equally well for such symmetric subrings, however we shall carry them out only for $\overline{S}$. In fact, our first result shows that $\overline{S}$ itself is a symmetric subring.

LEMMA 6.5.1. For any $x \in R$, $x\overline{S}x^* \subset \overline{S}$.

Proof. Clearly $xSx^* \subset S$. We also know by Example 2 following Theorem 2.1.4, that $\overline{S}$ is a Lie ideal of $R$. We prove the result by showing $xs_1s_2\cdots s_nx^* \in \overline{S}$ for $s_1, \ldots, s_n \in S$; the proof is by induction on $n$.

Now

$$\begin{aligned} xs_1\cdots s_nx^* = {} & (xs_1\cdots s_{n-1} - s_1\cdots s_{n-1}x)(s_nx^* - x^*s_n) \\ & + s_1s_2\cdots s_{n-1}xs_nx^* - s_1\cdots s_{n-1}xx^*s_n \\ & + xs_1\cdots s_{n-1}x^*s_n \; . \end{aligned}$$

Since $xx^* \in S$, $xs_1\cdots s_{n-1}x^* \in \overline{S}$ (by induction) and $\overline{S}$ is a Lie ideal of $R$, we see that $xs_1 \cdots s_nx^* \in \overline{S}$. This proves the lemma.

We now can prove the first of these kinds of theorems intertwining $\overline{S}$ and R.

THEOREM 6.5.1. If R is semi-prime then $\overline{S}$ is semi-prime.

Proof. Suppose that $a \in \overline{S}$ is such that $a^2 = 0$ and $a\overline{S}a = 0$. We want to show that $a = 0$; this will prove the semi-primeness of $\overline{S}$.

Since $\overline{S}$ is a Lie ideal of R, if $x \in R$ then $axa = a(xa) - (xa)a \in \overline{S}$. Hence, by Lemma 6.5.1, for any $y \in R$, $yaRay^* \subset y\overline{S}y^* \subset \overline{S}$, whence $ayaRay^*a \subset a\overline{S}a = 0$. However, since $y + y^* \in S \subset \overline{S}$, $a(y+y^*)a = 0$, whence $aya = -ay^*a$. Thus $ayaRaya = ayaRay^*a = 0$ for any $y \in R$. Because R is semi-prime, we get $aya = 0$ for all $y \in R$, hence $a = 0$. This proves the theorem.

Theorem 6.5.1 allows us to derive easily several results. We include them in the next few theorems.

THEOREM 6.5.2. Let R be any ring with involution. Then

1. if R has no nil ideals, $\overline{S}$ has no nil ideals,

2. if R has no locally nilpotent ideals, $\overline{S}$ has no locally nilpotent ideals.

Proof. We prove both results at the same time. Under either assumption R is semi-prime, hence $\overline{S}$ is semi-prime by Theorem 6.5.1.

If $\overline{S}$ is commutative, being semi-prime it has no nilpotent elements and the theorem is correct for $\overline{S}$ in that siutation. If $\overline{S}$ is not commutative, by Theorem 2.1.5, $\overline{S}$ has a non-zero ideal M of R and $M \supset [\overline{S}, \overline{S}]$. If $W \neq 0$ is a nil ideal, or a locally nilpotent ideal of $\overline{S}$ then $MWM \subset W$ is a nil or locally nilpotent ideal of R, hence is 0. Because R is semi-prime, from $MWM = 0$ we get $MW = 0$ and so

$M \cap W = 0$. If $a \in \overline{S}$, $b \in W$ then

$$ab - ba \in [\overline{S}, \overline{S}] \cap W \subset M \cap W = 0 ;$$

therefore $b \in Z(\overline{S})$. Since $b \in W$, $b$ is nilpotent; however, $\overline{S}$ is semi-prime and $b$ is nilpotent and in $Z(\overline{S})$. In consequence, $b = 0$, and so $W = 0$ contrary to $W \neq 0$. With this the theorem is proved.

We now prove a simple, formal result.

LEMMA 6.5.2. Suppose $u \in R$ is such that $u$ commutes with all $x + x^*, xx^*, xux^*$ and $ux - xu$ for every $x \in R$. Then $(ux - xu)^2 = 0$ for all $x \in R$.

<u>Proof.</u> From $u(x+x^*) = (x+x^*)u$ we have $ux - xu = -(ux^* - x^*u)$. By this result and the Jacobi identity, $uxx^* - xx^*u = 0$ gives us $(ux - xu)x^* = x(ux - xu)$. Finally, $uxux^* - xux^*u = 0$ gives

$$\begin{aligned} 0 = (ux - xu)ux^* + xu(ux^* - x^*u) &= u(ux - xu)x^* - xu(ux - xu) \\ &= ux(ux - xu) - xu(ux - xu) \\ &= (ux - xu)^2 . \end{aligned}$$

An immediate consequence of this is

THEOREM 6.5.3. If $R$ is semi-prime, then $Z(\overline{S}) \subset Z(R)$.

<u>Proof.</u> Since $\overline{S}$ is a Lie ideal of $R$, trivially $Z(\overline{S})$ is a Lie ideal of $R$. If $u \in Z(\overline{S})$, $u$ therefore commutes with all $x+x^*, xx^*$ and $ux - xu$ for every $x \in R$. Moreover, by Lemma 6.5.1, $xux^* \in \overline{S}$ so it, too, commutes with $u$. By Lemma 6.5.2, $(ux - xu)^2 = 0$. However, $ux - xu \in Z(\overline{S})$ and, by Theorem 6.5.1, $\overline{S}$ is semi-prime. The upshot is that $ux - xu = 0$, whence $u \in Z(R)$.

We can now tie together algebraicity properties for $\overline{S}$ with such properties of $R$.

THEOREM 6.5.4. If $R$ is an algebra with involution over a field $F$, then

1. if $\overline{S}$ is algebraic, so is $R$.

2. if $\overline{S}$ is locally finite, so is $R$.

<u>Proof.</u> We carry out the proof of the two parts simultaneously. Let $A$ be the algebraic kernel (i.e., the sum of all algebraic ideals of $R$) or the locally finite kernel (i.e., the sum of all locally finite ideals of $R$) of $R$ according to which hypothesis of the theorem we are using. Then $A = A^*$, $R/A$ is semi-prime and is without algebraic or locally finite non-zero ideals. Moreover, since the ideal generated by $[\overline{S}, \overline{S}]$ is in $A$ — by our hypothesis on $\overline{S}$ and the fact that $R[\overline{S}, \overline{S}]R \subset \overline{S}$ — if $s \in S$ maps onto $u$ in $R/A$ then $u$ commutes with all $y + y^*$, $yy^*$, and $yuy^*$ and $uy - yu$ in $R/A$. Hence, by Lemma 6.5.2, $(uy - yu)^2 = 0$ for all $y \in R/A$. Also, if $x \in R$ maps on $y \in R/A$, then $sx - xs \in \overline{S}$ so, for any $z \in R/A$, we get, if $v = uy - yu$ that $v$ commutes with $z + z^*$, $zz^*, zvz^*$ and $vz - zv$. Thus $(vz - zv)^2 = 0$. Because $v^2 = 0$, we get $(vz)^3 = 0$. But then $R/A$ has a nilpotent ideal, unless $v = 0$. We therefore have that $v = 0$ and so $u \in Z(R/A)$. In other words, every image of a symmetric element in $R$ is central in $R/A$.

If $w \in R/A$ then $w + w^*$ and $w^*w$ are images of symmetric elements of $R$, hence are central in $R/A$ and are algebraic over $F$. But $w^2 - (w + w^*)w + w^*w = 0$, hence $w$ is algebraic over $F$. Moreover, since $R/A$ is quadratic over $Z(R/A)$, $R/A$ satisfies a polynomial identity. Thus $R/A$ is locally finite (see Theorem 6.4.3 in [10]). Consequently $R/A = 0$ and so $R = A$, proving the theorem.

With techniques like the ones we have used it is easy to show that if R is *-simple or *-prime then so is $\overline{S}$. We shall not do it here but leave it to the reader. Likewise, if R is prime, then so is $\overline{S}$. We do prove one final result of this nature:

THEOREM 6.5.5. If R is semi-simple, then so is $\overline{S}$.

Proof. Suppose that $a \in J(\overline{S})$; if $x \in R$ then $ax + ax^* + axax^* = a(x + x^* + xax^*)$ is in $J(\overline{S})\overline{S} \subset J(\overline{S})$, hence is quasi-regular. This tells us — from the associativity of the operation $u + v + uv$ — that $ax$ is quasi-regular for all $x \in R$. But then $a \in J(R) = 0$. Hence $J(\overline{S}) = 0$.

The proof can easily be modified to show the

COROLLARY. $J(\overline{S}) = \overline{S} \cap J(R)$.

We leave a few more remarks to the reader for proof:

1. if R is primitive or *-primitive, then so is $\overline{S}$
2. if R is von Neumann regular, then so is $\overline{S}$
3. if R is semi-simple artinian, then so is $\overline{S}$.

We prove one theorem involving chain conditions

THEOREM 6.5.6. If R is a semi-prime Goldie ring, then so is $\overline{S}$.

Proof. Since the ascending chain conditions on annihilators needed in the definition of a Goldie ring are inherited by subrings, they hold for $\overline{S}$. To show, therefore, that $\overline{S}$ is Goldie we only must verify that it has no infinite direct sum of right ideals.

If $\overline{S}$ is commutative, by Theorem 6.5.3 $\overline{S} \subset Z$, in which case the above remark is trivial. So we may suppose that $[\overline{S}, \overline{S}] \neq 0$, in which case $\overline{S} \supset R[\overline{S}, \overline{S}]R = I$. Suppose that $\sum_{\alpha} \rho_{\alpha}$ is a direct sum of right ideals of $\overline{S}$; then $\sum_{\alpha} \rho_{\alpha} I$ is a direct sum of right ideals of R. Since R is Goldie, $\rho_{\alpha} I = 0$ for all but a finite number of $\alpha$. Since $I \supset [\overline{S}, \overline{S}]$, we get that $\rho_{\alpha}$ is commutative, hence by Lemma 1.1.5, $\rho_{\alpha} \subset Z(\overline{S}) \subset Z$ for all but a finite number of $\alpha$. But then $\rho_{\alpha} = 0$ for all but a finite number of $\alpha$, whence $\overline{S}$ is Goldie.

There are some converse theorems — not precise converses, but close — to the theorems above. By this we mean, if $\overline{S}$ satisfies a certain condition then so does R, provided R is decent enough. We refer the reader to the papers by Lanski and Lee cited earlier for the exact results and their details.

We shall not go into as much detail about the relative relationship of $\overline{K}$ and R as we did for $\overline{S}$ and R. We shall merely content ourselves with some sample results. The reader who wants to know more about this area can look in [14]. We first show that if R is semi-prime then $\overline{K}$ must be semi-prime. Our first step towards this goal is

LEMMA 6.5.3. If R is semi-prime and $a^* = -a$ is such that $a\overline{K}a = 0$ then $a = 0$.

<u>Proof.</u> If $x \in R$ then $x - x^* \in K$, hence $a(x - x^*)a = 0$, that is to say, $axa = ax^*a$. Also, $xax^* \in K$, hence $a(xax^*)a = 0$. These results together give $axaxa = 0$. Thus $aR$ is nil of index of nilpotency 3; by Lemma 2.1.1, $a = 0$.

We next show that the center of $\overline{K}$ is semi-prime when R is.

LEMMA 6.5.4. If R is semi-prime then $Z(\overline{K})$ has no non-zero nilpotent element.

Proof. If $a \neq 0$ in $Z(\overline{K})$ has square 0, then, by Lemma 6.5.3, a cannot be skew. Hence if $Z(\overline{K})$ has nilpotent elements it has an element $b \neq 0$, $b^* = b$, such that $b^2 = 0$. If $k \in K$ then bk is skew and $bk\overline{K}bk = b^2k\overline{K}k = 0$, whence, by Lemma 6.5.3, $bk = 0$. If $x \in R$ then $xb - bx^* \in K$, therefore $b(xb - bx^*) = 0$; since $b^2 = 0$ this yields that $bxb = 0$ for all $x \in R$. Because R is semi-prime we get $b = 0$, a contradiction.

We are now ready to prove

THEOREM 6.5.7. If R is semi-prime then so is $\overline{K}$.

Proof. Suppose that $\overline{K}$ is not semi-prime; then it is immediate that $\overline{K}$ has an ideal $W \neq 0$ with $W^* = W$ and $W^2 = 0$. If $a \in W$ then $a\overline{K}a \subset W^2 = 0$; by Lemma 6.5.3, no element in W can be skew, hence, since $W^* = W$, every element in W is symmetric.

If $a \in W$, $u \in \overline{K}$ then $au \in W$ consequently $au = (au)^* = u^*a$. Therefore, if $u, v \in \overline{K}$, $auv = (uv)^*a = v^*u^*a = v^*au = avu$ and so $a[\overline{K}, \overline{K}] = 0$. In particular, if $x \in R$ and $k \in [K, K]$, $a(x - x^*)k = 0$, and so $axk = ax^*k$.

If $k \in [K, K]$, $x \in R$ then $xkx^* \in K$ hence $a(kxkx^* - xkx^*k) = 0$; since $ak = 0$ we get from this

(1) $axkx^*k = 0$ for all $x \in R$, $k \in [K, K]$

Applying * to (1) yields

(2) $kx^*kxa = 0$ for all $x \in R$, $k \in [K, K]$.

Linearize (1) on $x$ to get $axky^*k + aykx^*k = 0$ for $x, y \in R$. Multiplying this on the right by $xa$ and using (2) gives us

(3) $axkRkxa = 0$, for $x \in R$, $k \in K$.

If $U = RaxkR$ and $V = RkxaR$ then $UV = 0$, and because $R$ is semi-prime, we have that $U \cap V = 0$. But

$$V \supset aVa = aRkxaRa = a(RkxaR)^* a = aRax^*kRa$$
$$= aRaxkRa = aUa$$

since we have $ax^*k = axk$. Therefore $aUa = aVa = 0$. Thus $aRaxkRa = 0$; $R$ being semi-prime, we conclude that $axk = 0$. Thus $aR[K, K] = 0$. Using the Jacobi identity we have $aR[\overline{K}, \overline{K}] = 0$.

If $A$ is the ideal of $R$ generated by $a$, the above tells us that $A \cap [\overline{K}, \overline{K}] = 0$. But, if $b \in \overline{K}$ then $ab - ba \in A \cap [\overline{K}, \overline{K}] = 0$. In short, $a$ must be in $Z(\overline{K})$. However, by Lemma 6.5.4 we know that $Z(\overline{K})$ has no nilpotent elements. Therefore $a = 0$ and $W = 0$. With this contradition we are done.

We conclude this section with another result in the same vein.

THEOREM 6.5.8. If $R$ has no nil ideals then $\overline{K}$ has no nil ideals.

<u>Proof.</u> $R$ is semi-prime, hence $\overline{K}$ is semi-prime by Theo rem Theorem 6.5.7. Suppose that $N \neq 0$ is a nil ideal of $\overline{K}$. We may suppose that $N^* = N$.

If $[K^2, K^2] = 0$ then, by Theorem 5.1.2, $R$ satisfies a polynomial identity, hence $\overline{K}$ must satisfy this polynomial identity. But a ring with P.I. which is semi-prime has no nil ideals. Therefore we may suppose that $[K^2, K^2] \neq 0$.

In that case $U = R[K^2, K^2]R \subset \overline{K^2} \subset \overline{K}$ since $[K^2, K^2]$ is a Lie ideal of R. But then $UNU \subset N$ is a nil ideal of R; in consequence, $UNU = 0$, and so $UN = 0$. Therefore $U \cap N = 0$. But then $N \cap [K^2, K^2] = 0$; since $N^* = N$ this tells us that N must satisfy a polynomial identity; for if $u, v \in N \cap K$ then $u^2v^2 - v^2u^2$ is in $N \cap [K^2, K^2] = 0$, whence by Theorem 5.1.2, N satisfies a P.I. But then N contains a nilpotent ideal; since N is an ideal of $\overline{K}$, $\overline{K}$ thus contains a nilpotent ideal. However $\overline{K}$ is semi-prime. The theorem is proved.

We should add that one gets the feeling that there must be some omnibus theorem inter-relating $\overline{S}, \overline{K}$ and R which would give the results proved here as special cases. This result is yet to be enunciated or found. The varied results we have proved here seem to give evidence that there is something far more general going on in this context than has come out in the particular results.

One might wonder if the relation between $\overline{S}$ and R, say, is that they are Morita equivalent. An example due to R. Snider smashes this hope. Let F be a field and let $U = \begin{pmatrix} 0 & 1 \\ -1 & 0 \end{pmatrix}$ ; on $F_{2n}$ define the involution * by

$$a^* = \begin{pmatrix} U^{-1} & & \\ & \ddots & \\ & & U^{-1} \end{pmatrix} a' \begin{pmatrix} U & & & \\ & U & & \\ & & \ddots & \\ & & & U \end{pmatrix}$$

where a' is transpose of a. Let R be the subring of the infinite matrices over F consisting of matrices of the form

$$\begin{pmatrix} A & & & & \\ & B & & O & \\ & & \ddots & & \\ & O & & B & \\ & & & & \ddots \end{pmatrix}$$

where $A$ is a $2n \times 2n$ matrix and $B$ is a $2 \times 2$ matrix of the form $\begin{pmatrix} \alpha & \beta \\ 0 & \gamma \end{pmatrix}$. The involution defined above induces an involution on $R$. If char $F \neq 2$ then the subring $\overline{S}$ generated by the symmetrics is the set of all matrices of the form

$$\begin{pmatrix} A & & & & & & \\ & a & & & & & \\ & & a & & & & \\ & & & \ddots & & & \\ & & & & a & & \\ & & & & & \ddots & \end{pmatrix}.$$

As can be verified, $\overline{S}$ and $R$ have lattices of two-sided ideals which are not isomorphic. But then $\overline{S}$ and $R$ cannot be Morita equivalent.

## 6. Finite Generation

In this final section of the book we develop some results proved recently by Osborn [16]. Their flavor and emphasis is somewhat different from that of the things that have preceded. However, they are in the general direction we have followed, that of intertwining the nature of $S$, or of $K$, and that of $R$ itself. The property that we look at here is finite generation, and how imposing finite generation on $R$ affects $S$, and to a lesser extent, $K$.

The first theorem is a very pretty result due to Osborn,

THEOREM 6.6.1. Let $A$ be a commutative ring such that $\frac{1}{2} \in A$, and let $R$ be an algebra with involution over $A$. If $R$ is finitely generated as an algebra over $A$ then $S$ is finitely generated as a Jordan algebra over $A$.

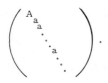

Proof. If $s \in S$ and $k \in K$ we use the notation $s' = [s,k] = sk - ks$, $s'' = ksk$, and $s''' = ksk^2 - k^2sk$. Of course, the elements $s'$, $s''$, and $s'''$ are all in $S$. We note some formal closure relations under commutation with $k$; here $a \circ b = ab + ba$. These are:

$$[s,k] = s', \quad [s',k] = s \circ k^2 - 2s'', \quad [s'',k] = s''',$$

and

$$[s''',k] = s'' \circ k^2 - (s \circ k^2) \circ k^2 + s \circ k^4 .$$

From these we have that the Jordan subalgebra of $S$ generated by $s, s', s'', s'''$, and $k^2$ is closed with respect to commutation by $k$.

A further bit of notation: if $C_0 = D \cup E$ where $D \subset S$ and $E \subset K$, and if $k \in K$, then $C_1 = \{s, s', s'', s''' \mid s \in D\} \cup E$ and we write $C_1 = < C_0, k>$ to denote that we have obtained $C_1$ from $C_0$ in the above manner. As in the paragraph above, the symmetric elements of $C_1$ generate a Jordan subalgebra of $S$ closed under commutation with $k$.

Since $R$ is a finitely generated A-algebra, and since $\frac{1}{2} \in A$, by using the symmetric and skew parts of the generators of $R$ we may assume that $R$ is generated by $B_0 = \{s_1, \ldots, s_q, k_1, \ldots, k_r\}$ with the $s_i \in S$ and the $k_j \in K$. Let $C_0 = B_0 \cup \{k_i k_j + k_j k_i \mid 1 \leq i, j \leq r\}$. Since $C_0 \supset B_0$, $C_0$ generates $R$.

Define the sets $C_i$, for $1 \leq i \leq r$, inductively by $C_i = <C_{i-1}, k_i>$. If $B$ is a subset of $S \cup K$ and if $x_1, \ldots, x_n \in B$ let $\{x_1 x_2 \cdots x_n\} = x_1 x_2 \cdots x_n + (-1)^j x_n \cdots x_2 x_1$, where $j$ is the number of $x_i$'s which are in $K$. Clearly as $n$ varies, the elements $\{x_1 \cdots x_n\}$ for $x_i \in B$ are in $S$ and span the Jordan subalgebra of symmetric elements in the associative subalgebra generated by $B$. Our aim is to show that the elements of $C_r$ together with certain elements $\{x_1, \ldots, x_n\}$, where the $x_i \in C_r$ and $n \leq r+3$, generate $S$ as a Jordan algebra.

Since $C_1 \supset C_0 \supset B_0$, $C_1$ generates $R$ hence $S$ is spanned over $A$ by the elements $\{x_1 \cdots x_n\}$ where the $x_i$ are in $C_1$. We assert that $S$ is spanned by the elements $\{x_1 \cdots x_n\}$ where the $x_i \in C_1 - \{k_1\}$ and the elements $\{k_1 x_2 \cdots x_n\}$ where the $x_i \in C_1 - \{k_1\}$. It is clearly enough to show that all $\{x_1 \cdots x_n\}$ with $x_i$ arbitrary in $C_1$ are so obtain obtainable.

Suppose that $\{x_1 \cdots x_n\}, x_i \in C_1$ is not obtainable in the span of the elements of the form described above, and that $n$ is minimal of this nature. Among all such, pick that one involving a minimal number of $x_i$'s equal to $k_1$. Finally, among all such elements, pick that $\{x_1 \cdots x_n\}$ in which the sum of the distances from the left of the $k_1$'s occuring is minimal. Suppose that $x_i = k_1$ but $x_j \neq k_1$ for $j > i$. If $x_{i-1} = k_1$ then $x_{i-1} x_i = k_1^2 \in C_1$ and we could replace the product $x_{i-1} x_i$ by one element of $C_1$, thus decreasing $n$; by our choice of $n$ we would have a contradiction. On the other hand, if $x_{i-1} = k_j$ for some $j \neq 1$ then $x_{i-1} x_i = k_j k_1 = (k_1 k_j + k_j k_1) - k_1 k_j$ and, substituting this in $\{x_1 \cdots x_n\}$ we get $\{x_1 \cdots x_n\}$ as a sum of two terms. In the first, since $k_1 k_j + k_j k_1 \in C_1$ we have lowered $n$, so it is obtainable in the desired form. In the second, involving $k_1 k_j$, $k_1$ has been moved one unit to the left, so by our ordering, it too is obtainable in the desired form. Finally, if $x_{i-1} = t \in C_1 \cap S$ then $x_{i-1} x_i = t k_1 = [t, k_1] + k_1 t$. The element $[t, k_1] = t'$ is in the Jordan algebra of symmetric elements generated by the symmetric elements of $C_1$, so expressing $t'$ in terms of these elements (none of which is $k_1$) we get an expression for

$\{x_1 \cdots x_{i-2} t' x_{i+1} \cdots x_n\}$ involving fewer $k_1$'s so this latter element is of the desired form. The element $k_1 t$ again gives a shift of one unit to the left by $k_1$, so by our ordering we are done. Putting these pieces together we have that here, too, in case $x_{i-1} = t \in C_1 \cap S$, the element $\{x_1 \cdots x_n\}$ is obtainable in the desired form. This proves our assertion.

We now assert that $S$ is spanned by the elements $\{x_1 \cdots x_n\}$ where $x_i \in C_2 - \{k_1, k_2\}$ and the elements $\{k_2, x_3, \ldots, x_n\}$, $x_i \in C_2 - \{k_1, k_2\}$ and the elements $\{k_1, x_2, \ldots, x_n\}$ where $x_2 \neq k_1$ and where $x_3, \ldots, x_n \in C_2 - \{k_1, k_2\}$. The argument is a refinement of the argument given above, and we leave it to the reader.

Continuing this way — weeding out one $k_i$ at a time — we get that $S$ is spanned by elements of the form $\{x_1 \cdots x_n\}$ where $x_i \in C_r \cap S$ and elements $\{k_{i_1} k_{i_2} \cdots k_{i_m} x_{m+1} \cdots x_n\}$ where $1 \leq i_1 < i_2 < \cdots < i_m \leq r$ and where $x_{m+1}, \ldots, x_n$ are in $C_r \cap S$. We shall now show that these elements are generated by Jordan multiplication by the tetrads $\{x_1 x_2 x_3 x_4\}$ with $x_i \in C_r \cap S$ and the elements $\{k_{i_1} k_{i_2} \cdots k_{i_m} x_{m+1} \cdots x_n\}$ where $1 \leq i_1 < i_2 < \ldots < i_m \leq r$ and $x_{m+1}, \ldots, x_n \in C_r \cap S$ <u>and</u> $n - m < 4$. If this is so, we are done for $m \leq r$ hence $n \leq r+3$ follows, and we would have that $S$ is finitely generated as a Jordan algebra.

Let $S'$ be the Jordan algebra generated by the elements $\{x_1 x_2 x_3 x_4\}$ with $x_i \in C_r$ and $\{k_{i_1} k_{i_2} \cdots k_{i_m} x_{m+1} \cdots x_n\}$ where $1 \leq i_1 < i_2 < \ldots < i_m \leq r$, $x_i \in C_r \cap S$ and $n - m < 4$. We want to show that $S' = S$. We follow the type of arguments given in the proof of Theorem 7.6 in Cohn's book [3]. Suppose that $S' \neq S$, and suppose that all $\{y_1, \ldots, y_k\}$, $y_i \in C_r$, with $k < n$ are in $S'$. Let $y_1, \ldots, y_n \in C_r$; we aim to show that $\{y_1 y_2 \cdots y_n\} \in S'$.

If $y_n \in C_r \cap S$ then

$$S' \ni \{y_1 \cdots y_{n-1}\} \circ y_n = \{y_1 \cdots y_n\} + \{y_n y_1 \cdots y_{n-1}\},$$

hence

$$\text{(1)} \qquad \{y_1 \cdots y_n\} \equiv -\{y_n y_1 \cdots y_{n-1}\} \mod S'$$

for all $y_1, \ldots, y_{n-1} \in C_r$, $y_n \in C_r \cap S$.

If $y_{n-1}$ is also in $C_r \cap S$ then

$$\begin{aligned} S' \ni \{y_1 \cdots y_{n-2}\} \circ \{y_{n-1} y_n\} &= \{y_1 \cdots y_{n-2} y_{n-1} y_n\} \\ &+ \{y_1 \cdots y_{n-2} y_n y_{n-1}\} + \{y_n y_{n-1} y_1 \cdots y_{n-2}\} \\ &+ \{y_{n-1} y_n y_1 \cdots y_{n-2}\} . \end{aligned}$$

Using (1) for $y_n$ and $y_{n-1}$ we get that

$$2\{y_1 \cdots y_{n-2} y_{n-1} y_n\} \equiv -2\{y_1 \cdots y_{n-2} y_n y_{n-1}\} \mod S'$$

and since $\frac{1}{2} \in A$, we end up with

$$\text{(2)} \qquad \{y_1 \cdots y_{n-2} y_{n-1} y_n\} \equiv -\{y_1 \cdots y_{n-2} y_n y_{n-1}\} \mod S'.$$

If, in addition, $y_{n-2}$ and $y_{n-3}$ are also in $C_r \cap S$, then $\{y_1 \cdots y_{n-4}\} \circ \{y_{n-3} y_{n-2} y_{n-1} y_n\}$ is in $S'$; but

$$\begin{aligned} \{y_1 \cdots y_{n-4}\} \ \{y_{n-3} y_{n-2} y_{n-1} y_n\} &= \{y_1 \cdots y_{n-4} y_{n-3} \cdots y_n\} \\ &+ \{y_1 \cdots y_{n-4} y_n \cdots y_{n-3}\} + \{y_{n-3} \cdots y_n y_1 \cdots y_{n-4}\} \\ &+ \{y_n \cdots y_{n-3} y_1 \cdots y_{n-4}\} . \end{aligned}$$

Using (1) and (2) then gives us that $4\{y_1 \cdots y_n\} \in S'$, and so $\{y_1 \cdots y_n\} \in S'$.

This shows that if the last 4 entries in $\{y_1 \cdots y_n\}$ are in $C_r \cap S$ then $\{y_1 \cdots y_n\} \in S'$. So we get if $\{y_1 \cdots y_n\} \notin S'$ then at most 3 of the last entries are in $C_r \cap S$; but then the element $\{y_1 \cdots y_n\}$ is of the second type amongst the generators of $S'$, so is in $S'$. This proves the theorem.

Let $R$ be any finitely generated algebra over a commutative ring $A$ such that $\frac{1}{2} \in A$. If $R^o$ is the opposite algebra of $R$ then $R \oplus R^o$ relative to the exchange involution is a ring with involution which is finitely generated over $A$. By Theorem 6.6.1, the Jordan algebra of symmetric elements in $R \oplus R^o$ is finitely generated as a Jordan algebra. Just reading off, from the form of a symmetric element in $R \oplus R^o$, the consequence of this we have

THEOREM 6.6.2. Let $R$ be a finitely generated algebra over a commutative ring $A$ such that $\frac{1}{2} \in A$. Then $R$, relative to the Jordan product $a \circ b = ab + ba$, is a finitely generated Jordan algebra.

It would be interesting to know whether or not the converses of these two theorems hold in fairly general circumstances, namely, if $R$ is semi-prime. (If $R$ is not semi-prime, by tacking on trivial direct summands in which you declare all elements to be skew gives trivial counter-examples.) Quite a bit of information, and a strong reduction of this problem, are to be found in Osborn's paper [16], but at the time of this writing the questions are open.

It is reasonable to ask if a similar story holds for the Lie algebra of skew elements in a finitely generated ring with involution. The answer is no. Let $F$ be a field of characteristic different from 2 and let R be the free algebra, $R = F[x, y]$, in the variables x and y. We impose an involution on R in which we declare x to be symmetric and y to be skew. Suppose that B is a finite set of skew elements of R that generate K as a Lie algebra. If $w = w_1 \cdots w_n$ is a word in x an y (so $w_1, \ldots, w_n$ are either x or y) then the elements $[w_1 \cdots w_n] = w_1 \cdots w_n - (-1)^j w_n \cdots w_1$, where j is the number of $w_i$'s equal to y, span K. We may assume that B consists of elements of this form, by using the elements of this form which express the elements of B as a linear combination.

Now, if two homogeneous elements w, z of R do not commute then then their commutator is a homogeneous element whose degree is the sum of the degrees of the two elements w and z. However, every homogeneous element of K is of degree at least 1, we get that the the commutator of any two elements of K is either 0 or a sum of homogeneous terms of degree at least 2. Thus any element of K which is of degree 1 in y must be a linear combination of elements of B of degree 1. Since this set contains all $yx^i + x^i y$ for all $i \geq 0$ and is infinite dimensional it cannot be spanned by B, which is finite. So, here, K cannot be finitely generated as a Lie algebra.

Using the same R above — however without involution — one can show by an argument similar to the one above that R, as a Lie algebra re the product $[a, b] = ab - ba$, is not finitely generated.

Bibliography

1. A. Z. Anan'in and E. M. Zyabko. On a question of Faith. Algebra i Logika, 13 (1974): 125-131, (English translation, 1975).

2. M. Chacron and I. N. Herstein. Powers of skew and symmetric elements in division rings. Houston Jour. Math. 1 (1975): 15-27.

3. P. M. Cohn. Universal Algebra. Harper and Row.

4. C. Faith. Algebraic division ring extensions. Proc. AMS 11 (1960): 45-53.

5. I. N. Herstein. A unitary version of the Brauer-Cartan-Hua theorem. Jour. Algebra 32 (1974): 555-560.

6. I. N. Herstein. On the hypercenter of a ring. Jour. Algebra 36 (1975): 151-157.

7. I. N. Herstein. A general commutativity theorem. (to appear).

8. I. N. Herstein. Certain submodules of simple rings with involution II. Canadian Math. Jour. XXVII (1975): 629-635.

9. I. N. Herstein. On a theorem of Brauer-Cartan-Hua type. Pacific Jour. Math. 57 (1975): 177-181.

10. I. N. Herstein. Non-Commutative Rings. Carus Monograph, 15, 1968.

11. I. Kaplansky. A theorem on division rings. Canadian Math. Jour. 3 (1951): 290-292.

12. C. Lanski. On the relationship of a ring and the subring generated by its symmetric elements. Pacific Jour. Math. 44 (1973): 581-592.

13. C. Lanski. Chain conditions in rings with involution. Jour. London Math. Soc. 9 (1974): 93-102.

14. P.H. Lee. On subrings of rings with involution, (to appear).

15. J.A. Loustau. Radical extensions of Jordan rings. Jour. Algebra 30 (1974): 1-11.

16. J.M. Osborn. Rings with involution and finite generation, (to appear).

17. D. Topping. On linear combinations of special operators. Jour. Algebra 10 (1968): 516-521.